Dedication

Because learning does not occur only in the morning ...
but especially for Cheri, Britni, and Brooke

Advanced Methods in Distance Education:
Applications and Practices for Educators, Administrators, and Learners

Table of Contents

Section I
Foundations of Instruction and Learning at a Distance

Section II
Adult Learning Theory

Section III
Systematic Instructional Design

Section IV
Technology Knowledge and Skills

Foreword

I love this book and recommend it highly! You might wonder why the president of the American Distance Education Association would associate herself with a book that focuses on distance education delivery systems. The answer is threefold: permanence, integration, and perspective.

Permanence

As a seasoned professional in the distance education field, I have seen a lot of books come and go since I finished my PhD in adult and continuing education in the early 1970s. And while I may be getting grouchy as I approach the big 60 mark, I am astounded by the shallowness of so much of what I have seen written in the past 10 years about learning via technology. Perhaps never has so much been written about so little. In fact, it may be a good thing that much of this body of "knowledge" will disappear because it neither relates to the past nor the future.

This book is clearly an exception. It is rich in content, demonstrates deep knowledge of the field, and connects the tried and true from the past to the as-yet unleashed promise of the future. It deserves a place of "permanence" in the field. It should appeal, as the title suggests, to educators, trainers, and learners everywhere, and whenever we need to remember and renew our understanding of learning methods and quality practice. This book should hold special appeal because the theories and practices described have proven useful in varied cultural settings and with diverse populations. Not many books can withstand the critical permanence test on these dimensions. *Advanced Methods in Distance Education* is both timely and timeless.

Integration

This important "I word" is increasingly discussed as the missing piece in a world a bit mucked up in specialization. The fact that educators, administrators, and learners are all allowed to appear in the title suggests that readers might discover in the text how programs can be "learner centered" as well as demonstrate "quality instructional design," Dooley, Lindner, and Dooley's gift for pulling together really useful material over a broad set of areas. It is particularly unusual to see such a successful integration of the best of the best from adult education and training.

Academics and trainers too often stake out opposing positions: When is training education? Can training ever be education? How can instructional objectives be of any use in self-directed learning? I could go on with the set of debating society issues, but suffice it to say, you will find a very thoughtful, even wise, convergence approach that many should find provoking and useful.

Perspective

And finally, I think the really unique and creative aspect of this book relates to the notion of perspective. This book is BOTH science and art. My husband, an architect and engineer, has a treasured book titled *The Art of the Engineer*. The ease and flow of this book is artful and authentic as it moves between and among theories and from theories to practice to learners and around again.

Webster's Dictionary defines "perspective" as "the interrelation in which a subject or its parts are mentally viewed" (i.e., places the issues in proper perspective). Just like beauty, quality is in the eye of the beholder. Because this volume integrates concepts so skillfully, it highlights some unique and uncommon views that should appeal to many regardless of place.

Distance is very much a matter of perspective. It amuses and frustrates me that people think that words such as "distributed education" or "blended learning" encompass and can replace distance education. Distance education has long meant that everybody everywhere should have access to quality learning opportunities. Distance education is not a campus or place-bound concept. But then, that is a matter of my perspective.

Janet Poley
President/CEO
American Distance Education Consortium

Preface

Books that focus on the technology applications of distance education are quickly outdated. This book is different because it focuses on learner-centered instructional design and sound principles of adult learning. It is our belief that it is not the media but the method that makes a difference in effective instruction delivered at a distance.

This book was written for those (1) in higher education settings serving as teachers or instructors, instructional designers, media professionals, and/or administrative leaders; (2) in corporate or industry settings, including the military and other government organizations that provide professional or organization development; and (3) students or lifelong learners seeking to join the profession or to increase understanding of the field of distance education. A broad reach, you might think, but are we really that different? We like and defend the idea of "inclusivity"—all of us learning together about the field of distance education.

From research to practice is the purpose of this book. The book is written for those who do not have a background in "teaching and learning"; therefore, we provide the theoretical and philosophical basis for designing instruction in general and then add the components that make teaching or training at a distance unique and special.

We include current research studies to serve as a theoretical foundation and provide numerous practical guidelines and examples for reflection and elaboration. You will find additional reading links, case studies, or content applications within boxed information in the chapters. These boxed areas include activities or questions for *Thought and Reflection* that can be used as an individual exercise to take a break from the reading and think about how you would respond. Instructors may want to use these boxed activities for course activities as well. There are *Internet Connection* boxes with links to additional reading or examples of the principles being addressed in the chapter. Every chapter concludes with an *Application Exercise*. If you participate in these exercises, you will develop a lesson at a distance—a skill set that can be transferred to an entire course or training program.

The book is divided into six sections: Foundations of Instruction and Learning at a Distance, Adult Learning Theory, Systematic Instructional Design, Technology Knowledge and Skills, Administrative Issues, and Future Directions.

Section I provides the foundation for distance instruction and learning. Chapter I is an introduction to designing and delivering courses and programs at a distance. We define and describe the concept of benchmarking competencies for distance education professionals and learners based upon prior research studies conducted by the authors. The framework introduced in this chapter for best practices will be included throughout the book to guide practitioners and promote learning.

The theme for Chapter II is critical issues for educators and trainers engaged in distance education programs. Distance education as an innovation sets the stage. We introduce globalization as a powerful concept and moving force that affects decision making related to instructional development and delivery at a distance. The need to provide accessible knowledge and learning objects worldwide is imminent. This chapter also highlights the "No Significant Difference Phenomenon" and some research examples in regard to good instructional practices.

Chapter III focuses on the philosophical foundations of learning theory or models of learning. Concepts of behaviorism, cognitive information processing, and situated cognition/social learning theory are emphasized. An understanding of theoretical models of learning is essential for the design and delivery of instruction at a distance.

Section II of the book focuses on the use of adult learning theory (andragogy). Incorporating adult learning principles into the design and delivery of distance courses will result in more meaningful learning. Chapter IV includes a review of adult learning principles and learner differences that can impact learner engagement and ultimate success in a distance course. Andragogy provides various definitions and assumptions about adult learners that guide and direct the instructional design process. Age and generation, Kolb's learning style inventory, cognitive styles and controls, and multiple intelligences are some of the learner differences or characteristics that are discussed.

Chapter V concentrates on principles for engaging learners and fostering self-directedness. Additional explanations on characteristics of learners, including temperament/personality, gender, attrition rate, responsibility, interactions/engagement, and quality, set the stage for the notion of self-directed learning and the roles of educators and learners in this process. This section of the book will help instructors foster deeper and more meaningful learning by taking into account a learner's unique background, experiences, knowledge, skills, abilities, self-directedness, and/or personal learning styles and values.

Section III of the book emphasizes systematic instructional design. Chapters in this section incorporate writing instructional objectives, techniques for gaining attention and motivating learners, strategies for engaging the learner actively, and methods for assessing learning authentically. Chapter VI provides an overview of the instructional design process. The focus is on student-centered, rather than instructor-centered, design. We explore the ADDIE model as a template and provide scaffolding for you to create your own design.

The focus of Chapter VII is writing instructional objectives. It is a bit like learning the ABCs. Objective writing is "literacy" in relation to the instructional design process. The focus is on student-centered, rather than instructor-centered, design. We will provide a template and mention the use of concept mapping as a strategy to create your own design.

Gagné's first event for instruction is gaining attention. Why is that important? Chapter VIII explores the use of icebreakers and openers as a first step in creating an active learning environment. Audience assessment, building rapport, review of previous content, and advanced organizers are just a few reasons to incorporate icebreakers and openers into every instructional sequence. Learning is a social process. We need interaction with others (instructor and learners) in addition to the course content. When the learner is actively engaged, retention is enhanced and satisfaction increases.

If the objective is written correctly, then measuring outcomes is much easier. Assessment should be a continuous process. Questioning techniques through mediated communication (e-mail, threaded discussion, chat) provide an indication of confusion or understanding, allowing the instructor to make changes in delivery. Traditional testing for verification and the challenges it can pose for distance education are included in Chapter IX, along with the use of rubrics and authentic assessment techniques to measure whether the instructional objectives have been met.

Section IV addresses technology knowledge and skills. Of particular importance is the need for instructors and designers to understand access variability with bandwidth and different software requirements. Moreover, instruction must be designed for multiplatform use and for future technology development.

Chapter X discusses potential delivery technologies to bridge the distance between and among learners and instructors. Examples include print, audio, audio and video, and computer tools. Chapter XI introduces the multimedia tools, including graphic design balance and purpose, audio, streaming media, animation, and simulation.

Section V of the book covers administrative and management issues. For most instructional leaders, this issue is not only foreign, but likely an area that causes great consternation! It is critical for instructors and learners, as well as instructional designers, to be aware of these important issues so that the course or program will run smoothly.

Chapter XII addresses the major concepts that instructors, learners, and administrators need to know when delivering training at a distance. These issues cover aspects of learner support services, technical support, copyright concerns, and institutional models for rewarding and recognizing instructors. Budget and funding models, such as categories of cost and income, funding strategies, staffing issues, creating collaborative partnerships, and operational procedures, are discussed in this chapter.

Evaluation is an area that most would agree is necessary for program planning and delivery; however, it is also one of the most neglected. Chapter XIII explores programmatic and course evaluation, providing numerous examples of various types of evaluation.

Section VI concludes the book and provides points for reflection and future directions. Distance education has evolved from correspondence schools of the 1800s to delivery of training via desktop videoconferencing and the Internet today. But what does the future hold for this field? What will be some of the major changes in distance education 10, 20, or even 50 years from now?

We chose in Chapter XIV to emphasize educational and technological trends that could impact distance learning. We also include some visions for the future of distance education from the perspective of the learner, university faculty, and an international training and development specialist. We must all remember that as professionals in the field, we will have an impact on what the future will be like. But this assumes that we will stay current and constantly look for ways to expand our reach and promote global learning. Please join us on this journey!

Acknowledgments

This book would not be possible without the help, advice, and guidance of several individuals. We are grateful to the following people who coauthored with us on various chapters and/or sections within chapters. Their expertise helped to fill gaps in our knowledge base and make the book stronger.

For *Section I: Foundations of Instructional and Learning at a Distance*, we would specifically like to thank Dr. Tim Murphy, Associate Professor and Assistant Department Head for Graduate Programs in Agricultural Education, for his contributions on learning theory; and Dr. James A. Buford, Jr., Division Director, Ellis-Harper Management and Adjunct Professor at Troy State University and Auburn University for his contributions on benchmarking competency-based training.

For *Section II: Adult Learning Theory*, we would like to thank Susan Wilson, graduate assistant and student in the Department of Agricultural Education at Texas A&M University for her contribution on learner differences.

For *Section III: Systematic Instructional Design*, we would like to thank Dr. Atsusi Hirumi, Associate Professor of Instructional Technology at the University of Central Florida, for his thorough discussion on learner-centered instructional design and interactions to promote engagement in distance education. We would also like to thank Drs. Barry Boyd, Assistant Professor in the Department of Agricultural Education at Texas A&M University, and Kathleen Kelsey, Associate Professor at Oklahoma State University, for their guidance on learner-centered assessment.

For *Section IV: Technology Knowledge and Skills*, we would like to thank Dr. Walt Magnussen, Director for Telecommunications at Texas A&M University, for his contribution on technology infrastructure. We would also like to thank Dr. Rhonda Blackburn, Lead IT Consultant at Texas A&M University's Instructional Technology Services, and Ms. Yakut Gazi, a doctoral candidate in the Department of Educational Psychology at Texas A&M University, for their contribution on multimedia design issues.

For *Section V: Administrative and Management Issues*, we would like to thank Dr. Kathleen Kelsey for her expertise on best practices in distance education evaluation.

For *Section VI: Future Directions*, we would like to thank Chehrazade Aboukinane, a doctoral student in the Department of Agricultural Education at Texas A&M University, for her contributions on educational and technological trends and the glossary and index.

We also needed a fresh set of eyes to read over and edit the narrative for content and style. We would like to thank Dr. James Christiansen, our mentor, colleague, and friend, for editing this book. We solicited vignettes on the future of distance education and would like to thank all who provided timely and thought-provoking comments. Dr. Lynn Jones, Associate Professor at Iowa State University, was the viewpoint chosen, along with Dr. James Christiansen, to represent this group.

Janet Poley, CEO of the American Distance Education Consortium (ADEC), has been a friend and colleague for years. We appreciate her insightful foreword and are pleased to add this to the book.

The artwork for the Making Connections, Internet Connections, Thought and Reflection, and Application Exercises was designed by Christine Stetter working at the Instructional Material Services at Texas A&M University.

Our Department Heads, Drs. Glen Shinn, Christine Townsend and Jim Scheurich, provided true leadership, support, and encouragement while we were writing this book. Without their leadership to create an environment that encourages scholar ship, it would have been difficult to complete this project. To our collective faculties, who have had to live with us while we worked on this book ... the next round is on us. We appreciate your support, encouragement, and friendship. You have made our work not only rewarding but fun. Thank you!

We would like to thank the following people for their various contributions: Meera Algaraja, Eric Cartrite, Margarett DeGange, Mike Farrow, Teri Gerst,

Kim Hays, Jim Hynes, Kevin Jackson, Bob Keeshan, Yan Li, Aggie Manwell-Jackson, Penny Pennington, Lance Richards, Robert "Corky" Sarvis and Jemima Yakah.

And finally, we would like to thank our families for their love and understanding, and for just putting up with us as we worked on this book. Their encouragement allows us to excel at our jobs and, more importantly, in life.

Section I

Foundations of Instruction and Learning at a Distance

This section focuses on the core competencies for teaching and learning at a distance. We introduce teaching and learning philosophies as a foundational component for effective instruction and learning. Core competencies needed for distance education, including adult learning, instructional design, technology knowledge and skills, and administrative issues are discussed in this section. Communication skills, such as facilitation and providing meaningful feedback, are integrated throughout.

Chapter I

An Introduction to Designing and Delivering Courses and Programs at a Distance

with
James Buford, Jr., Ellis-Harper Management,
Troy State University and Auburn University, USA

 Making Connections

Meeting the needs of today's learners requires instructors and administrators to rethink delivery strategies and instructional methods. Many organizations are turning to distance education, because of its effectiveness, to help learners develop and improve their knowledge, skills, and abilities. The fact that you are reading this book leads us to assume that you have several questions about designing and delivering distance courses and programs. Our own experiences, research, and education have guided us in writing this book. We continue to question "What are the best practices for distance education?" "How do I assess meaningful learning?" and "How do I actively engage learners?"

Introduction

Distance education occurs when the instructor and learners are separated by location and/or time (Lindner & Murphy, 2001). This separation requires communication channels to bridge the "distance" between and among learners and instructors. This type of education is particularly appealing to learners whose responsibilities do not allow them to take advantage of traditional classroom or face-to-face training methods. Designing and delivering programs at a distance *is* different from face-to-face instruction, and those differences will serve as the focus of this book.

Distance education is not a new phenomenon. It was introduced in the United States in the late 1800s by correspondence study through the postal service, yet the theoretical foundations of distance education did not appear in scholarly writing until 1987 (Saba, 2003). The focus of this book is on distance education since the mid-1990s, when the integration of the World Wide Web and interactive video began to increase the visibility and usability of telecommunications through fiber optics, Internet Protocol (IP) addresses, and cable modems.

Many people use the terms "Internet" and "World Wide Web" interchangeably, but they are different. The Internet is a massive network of networks that connects computers together. The Internet is used for e-mail, newsgroups, and instant messaging. The Web is a way of accessing information over the Internet. The Web uses browsers to access documents using hyperlinks. Web documents contain graphics, sound, text, and video (Webopedia, 2002). Distance education, the Internet, and the World Wide Web have been defined operationally, but who are the distance learners involved, in what settings are they found, and where are growth areas expected?

Distance Learners and Distance Learning Settings

Who are distance learners? Typically, distance learners have been people who are adults with families, full- and part-time employees, living in rural areas, unable to afford full-time study, or military personnel. However, the fastest growing groups of distance learners are resident, on-campus students who want the flexibility to take courses that are not bound by time or place. Most of these students are enrolled at large, public institutions even though the largest percentage of growth is anticipated for schools in the for-profit sector (Allen & Seaman, 2003).

The largest university in the world is Anadolu University in Turkey with over 500,000 distance learners (MacWilliams, 2000). Most of these learners are working adults who hold full- or part-time jobs. The Open University of Hong Kong provides courses and programs to over 100,000 students. Spain's Universidad Nacional de Educación a Distancia has approximately 130,000 learners. Within the European Union, a variety of delivery strategies has been used, from traditional correspondence courses to computer conferencing, to two-way audio and video virtual classrooms. Through all these technologies, distance education and training within the European Union are expected to play a significant role in continuing to reach distance learners (Simonson, Smaldion, Albright, & Zvacek, 2003).

In a report from the Sloan Consortium, it was determined that over 1.6 million students in the United States took at least one online course during the fall of 2002. Over one third of these students took all of their courses online. This number was expected to increase by about 20% to be approximately 2 million students (Allen & Seaman, 2003). Not only are students willing and interested in learning online, but institutions and faculty are also embracing this delivery method. Online degree programs are now offered by 34% of higher education institutions in the United States (Allen & Seaman, 2003). "Major organizational changes and new developments in higher education are being accelerated by dynamic advances in global digital communications and increasingly sophisticated learning technologies ... Barriers to accessing learning opportunities are being reduced globally because of improved learning technologies" (Hanna, 1999, p. 19).

Distance Education Competencies

Distance education continues to expand because of growth of the Internet, increased information technology competencies, and reduced barriers to accessing and using the Internet (Lindner, 1999). Effective strategies for online training and instruction should focus on building virtual learning communities, making the technologies used to mediate communication as seamless and transparent as possible in order to efficiently and effectively reach as many learners as possible. Traditional barriers to distance education still exist, such as a lack of instructor and learner professional development and support, copyright and intellectual property issues, and few financial models to create and sustain distance education delivery, but these pose less of a barrier than attitudinal issues in promoting delivery of distance education. Rather than focusing on barriers, today's distance education professional should focus on the competencies to be successful in creating a learning community.

Courses and programs being delivered at a distance require a unique set of professional competencies for both instructors and administrators. Whether using synchronous or asynchronous methods of instruction, systematic instructional design can help stimulate motivation, increase interaction and social presence, and authenticate learning outcomes (Dooley, Lindner, & Richards, 2003). Principles of adult learning, including self-directed and student-centered learning approaches, are emphasized.

Benchmarking Competency-Based Training

A learner must possess certain knowledge, skills, and abilities in order to complete a planned instructional sequence and be successful (Lindner, Dooley, & Murphy, 2001). Knowledge is a body of information applied directly to the performance of a given activity. Skill is an observable competence to perform a learned act that requires motor ability (psychomotor). Ability is a present competence to perform an observable behavior or a behavior that results in an observable product. Competencies, therefore, establish the behavior requirements needed to be successful in any given learning environment. Buford and Lindner (2002) defined competencies as a group of related knowledge, skills, and abilities that affect a major part of an activity. Competency models can be used as recruitment and selection tools, as assessment tools, as tools to develop content and other instructional materials, as coaching, counseling, and mentoring tools, as career development tools, and as behavioral requirement benchmarking tools (Yeung, Woolcock, & Sullivan, 1996).

Based on a competency model developed by the American Society for Training and Development (ASTD), Thach and Murphy (1995) identified roles, outputs, and competencies of distance learning professionals within the United States and Canada. Their top-ten competencies portray the dual importance of both communication and technical skills in distance learning. These competencies in rank order are as follows: (1) interpersonal communication, (2) planning,

(3) collaboration/teamwork, (4) English (or language) proficiency, (5) writing, (6) organizational skills, (7) feedback, (8) knowledge of the distance learning field, (9) basic technology knowledge, and (10) technology access knowledge (Thach & Murphy, 1995). Williams (2000, 2003) replicated this study with similar results. Others have built complete degree programs (Ally & Coldeway, 1999) or certification programs to provide the coursework or professional development (competence) to work in the growing field of distance education.

Egan and Akdere (2004) explored roles and competencies from the perspective of graduate students specializing in distance education and compared them to previous works (Thach, 1994; Williams, 2003). They determined that agreement existed among and between experts but differed in prioritization of the competencies. Graduate students believed that technology competencies were most important.

Determined in a previous study by Dooley and Lindner (2002), the difficult task is in trying to measure and verify competence in a given profession. Industries as well as universities are struggling with appropriate techniques to document professional growth and learning over time. One method for addressing this problem is to develop and use competency-based and behaviorally anchored rating scales to measure learning. Behavioral anchors are defined as characteristics of core competencies associated with the mastery of content. Competency-based behavioral anchors are defined as performance capabilities needed to demonstrate knowledge, skill, and ability (competency) acquisition. Competency-based behavioral anchors require considerable time and effort to develop; however, they provide more accurate judgments than item-based scales (Buford & Lindner, 2002). Further, such anchors provide instructors and other expert raters with behavioral information useful in providing assessments and feedback to learners. Such information can help learners better understand their unique bundles of competencies and increase satisfaction, motivation, learning, and ultimately success (Drawbaugh, 1972). Competency-based feedback based on behaviors can provide a foundation for learner-centered instruction. Behavioral anchors can also be used to describe minimally acceptable knowledge, skills, and abilities on identified core competencies, thus giving instructors tools and information needed to improve instructional materials, evaluation processes, and delivery methods.

We have clustered the distance education competencies into six major areas or core competencies needed by learners and practitioners: adult learning theory, technological knowledge/skill, instructional design, communications skills, graphic design, and administrative issues (Table 1). For each core competency, behavioral anchors are presented as examples.

Internet Connections <http:www.cdlr.tamu.edu/>

The Center for Distance Learning Research is a public/private partnership between Texas A&M University and Verizon Corporation. In a continuing effort to disseminate information about the field of distance education, this Center offers a variety of certification programs. Certificates can be awarded in the areas of distance administration, technology integration for classroom teachers, virtual instructor, web publisher, technology coordinator, fiscal management of technology systems, collaborative videoconferencing, PC academy for high school students, and an overview of distance education.

Based on a scale of 1, being novice, and 7, being expert, we developed competency-based behavioral anchors at scale levels 2, 4, and 6 to authenticate ratings and standardize judgments of expert raters (Smith & Kendall, 1963) in order to benchmark learning (Table 2). The scale determines the

Table 1. Core competency behavioral anchors

Core Competency	Behavioral Anchors
Adult Learning Theory	• Philosophy of Teaching • Adult Learner Characteristics • Learning Styles
Technological Knowledge	• Web-Based Course Tools • Interactive Videoconferencing • Computer Hardware/Software • Communication Tools
Instructional Design	• Course Planning and Organization • Gaining Attention • Writing Instructional Objectives • Active Learning Strategies • Assessment and Evaluation
Communication Skills	• "Presenting" Content • Questioning and Facilitation • Feedback • Collaboration/Teamwork
Graphic Design	• Formatting Visuals • Interface Design • Multimedia Components
Administrative Issues	• Support Services • Copyright/Intellectual Property • Technology Access • Financial Considerations

learner's perceived competence, which can be measured throughout a course or training program to determine growth (learning) over time.

In the original Thach study (1994), 18 roles and outcomes (or instructor behaviors) for distance instructors were identified. These were based upon the core competency areas. To conclude this discussion of core competencies

Table 2. Competency-based behavioral anchors

Core Competency	Level	Competency-Based Behavioral Anchors
Adult Learning Theory	2	• Show someone how to do a literature review on student-centered learning
	4	• Present a short workshop on the theory of andragogy (adult learning)
	6	• Develop and deliver a student-centered training program that incorporates adult learner characteristics and student learning styles
Technological Knowledge/Skill	2	• Show someone how to log onto a computer and search the Internet
	4	• Show someone how to access and use Web course tools
	6	• Show someone how to design and execute a Web-delivered course using Web course tools
Instructional Design	2	• Use an icebreaker or opening to gain attention
	4	• Prepare a lesson plan
	6	• Write measurable instructional objectives for a curriculum that provides for student-centered learning
Communication Skills	2	• Facilitate a videoconference
	4	• Create virtual teams for discussion threads
	6	• Design appropriate synchronous and asynchronous communications methods for delivering course materials at a distance
Graphic Design	2	• Rely on technical experts to develop multimedia
	4	• Show someone how to develop a PowerPoint presentation with graphics
	6	• Show someone how to use animation, video streaming, and text to effectively deliver content
Administrative Issues	2	• Rely on technical experts for scheduling and copyright clearance
	4	• Identify and use available support services to plan and organize a course
	6	• Determine fiscal, human, and technical needs to plan and implement a curriculum entirely at a distance

Table 3. Checklist of suggested instructor behaviors for distance learning (DL)

Distance Learning Roles and Outcomes	Have You . . .
1. Planning Skills	• Determined your course goals? • Found out who are your students?
2. Instructional Design	• Developed instructional objectives? • Designed and provided materials to the learners that meet the instructional objectives?
3. Content Knowledge	• Incorporated current literature and practice?
4. Modeling of Behavior Skills	• Prepared your instructional materials following best practices? • Demonstrated skill with use of delivery technologies?
5. Interpersonal Communication	• Developed a relationship with each of your students? • Used examples relevant to the learners?
6. Feedback Skill	• Corresponded in a timely manner on assignments and questions?
7. Presentation Skills	• Incorporated a variety of methods? • Used icebreakers/openers to gain attention?
8. Technology Knowledge	• Learned the use and manage course delivery tools? • Developed redundancy in service or a backup plan?
9. Evaluation Skills	• Determined how you will access learning and inform the learners about their progress?
10. Collaboration/Teamwork	• Asked for a peer review of your materials and strategies? • Developed mechanisms for learners to work in virtual teams or communicate with you and other learners as needed?
11. Teaching Strategies	• Modified your instruction for distance learning? • Incorporated a facilitation role rather than an information giver role?
12. Facilitation and Group Process	• Promoted interaction between and among learners?
13. Needs Assessment	• Developed an understanding of how DL might impact these learners? • Accounted for learner differences?
14. Questioning Skills	• Developed higher order questioning strategies? • Synthesized questions to promote interaction and discussion?
15. Learning Style and Theory	• Developed an understanding of learning theory and incorporated it into the design and learning activities of your course?
16. Adult Learning Theory	• Provided opportunities for application of the material taught, which relates to the learners' experiences?
17. Advising/Counseling	• Communicated about ways to contact you? • Communicated your availability?
18. Support Service Knowledge	• Determined what resources are available to you (people, places, things)? • Determined what resources are available to your learners?

needed by distance learners and practitioners, we have provided a checklist of these roles in Table 3. The roles and outcomes are in the same order as the original Thach study and do not imply that they would occur sequentially.

After reading this text and participating in the application exercises, you will be able to measure your own learning in regard to proficiency based upon these competencies, roles, and outcomes.

Best Practices in Distance Education

When considering how to design and deliver effectively distance education courses and programs, *Principles of Good Practice* should be your guide. The American Distance Education Consortium (ADEC) offers four principles:

- Design for active and effective learning by considering the needs and characteristics of the learners, nature of the content, appropriate instructional strategies and technologies, and desired learning outcomes.

- Support the needs of the learner by providing advising, technical, and library support.

- Develop and maintain technological and human infrastructure.

- Sustain administrative and organizational commitment to quality by integrating distance education into the mission, providing financial commitment, encouraging faculty development and rewards, training to support those involved, and including marketing and management structures to promote and sustain distance education programs.

Initially drafted by the Western Cooperative for Educational Telecommunications and further developed by the Council of Regional Accrediting Commissions, the *Best Practices for Electronically Offered Degree and Certificate Programs* promotes well-established essentials of institutional quality and espouses the philosophy that virtual learning communities promote excellence in education (2001). Electronically offered programs both support and extend the roles of the educational institutions, with growing implications for institutional infrastructure. Methods change, but standards of quality endure. Instructor roles are becoming increasingly diverse and reorganized. Institutions and businesses should conduct sustained, evidence-based, and participatory inquiry as to whether distance learning programs are achieving targeted or proposed objectives.

 Internet Connections <http://www.adec.edu>

The American Distance Education Consortium (A*DEC) is an international consortium of state and land grant institutions providing economical distance education programs and services via the latest and most appropriate information technologies.

The Institute for Higher Education Policy prepared a report called *Quality on the Line: Benchmarks for Success in Internet-Based Distance Education* (NEA, 2000). The final report contains 24 benchmarks essential to quality, intended to assist policy makers. The primary components are listed below:

- A technology plan that includes security measures, reliability of technology, and centralized support
- Guidelines for course development, design, and delivery, including review of instructional materials to ensure they meet program standards
- An environment to stimulate student interaction with faculty and other students
- Advising about self-motivation and access to technology requirements as part of an orientation or preassessment phase prior to the start of a program
- Admission requirements, tuition and fees, books and supplies, technical and proctoring requirements, and student support services, as well as training and assistance on how to access these services
- Technical assistance in course development and continued support for faculty to ensure that the transition to online instruction is effective
- An evaluation plan and strategy to determine program effectiveness.

It is evident that all these entities have the same fundamental ideas in mind: creating a learning environment that provides stimulating learning experiences using communication technologies to connect learners and instructors.

Thought and Reflection

The *Principles of Good Practice* provides an outline for developing, delivering, or evaluating the quality of technology-assisted instruction (THECB, 1999). These *Principles* were adopted by the Texas Higher Education Coordinating Board. A self-study should focus on meeting quality standards.

It is certainly not necessary for you to provide a positive response to each of the questions below. These are suggestions appropriate to a wide range of educational programs. However, these questions should indicate course attributes that you may want to add, areas that you may wish to improve or further develop, and issues to be addressed in your course/program.

Certain assumptions are central to the *Principles of Good Practice* as well as this self-study:

1. The program or course offered electronically is provided by or through an institution that is accredited by an accrediting agency and authorized to operate where the program or course originates.
2. The institution's programs and courses holding specialized accreditation meet the same requirements when offered electronically.
3. The "institution" may be a single institution or a consortium of such institutions.
4. These principles are generally applicable to degree or certificate programs and to courses offered for academic credit.
5. It is the institution's responsibility to review the educational programs and courses it provides electronically and certify continued compliance with these principles.
6. Institutions offering programs or for-credit courses are responsible for satisfying all approval and accreditation requirements before students are enrolled.

Conclusions

Going back to our opening questions, what are the best practices to use in distance education? The Internet Connection sections and resources in the following chapters help to build the necessary foundation to design and deliver distance education courses and programs with *Principles of Good Practice* as your guide. In the following Application Exercise, you have the opportunity to explore this further. How do you assess meaningful learning? Although many assessment techniques can be used to measure learning, we suggest the use of authentic assessment tools, which are discussed more fully in chapter IX. More and more educational programs are using competency-based instructional modules to allow learners, particularly adults, the freedom to learn on demand the content and skills that are relevant to the current situation. Instructors can build individualized learning sequences by measuring competence at the beginning and end of the course or program to determine growth as a result of training. How do you actively engage learners? Interaction with the content, other learners, and the instructor is fundamental to applying best practices in distance education. Although we include interactive strategies for building virtual learning communities throughout this book, that is the emphasis of chapter IX.

Instructors and trainers are still struggling with the notion of quality and a belief often expressed that distance education is not as good as face-to-face instruction. That simply is not true, although teaching at a distance does require a unique set of competencies in order to create the social presence and interaction that is necessary for students to feel actively engaged in the learning process. This is the focus of the next chapter.

Distance education causes us to embrace change and creatively design new modes of delivery to respond to the needs of the learners. Attitudinal issues, such as how people perceive and react to distance education technologies, are far more important than structural and technical obstacles in influencing the use of distance education (McNeil, 1990). We hope that you are ready to try some new things and explore effective distance education practices and applications.

 Application Exercise

Search the WWW for courses or training programs online and jot down your initial thoughts on whether best practices were implemented.

References

Allen, I.E., & Seaman, J. (2003). *Sizing the opportunity: The quality and extent of online education in the United States, 2002 and 2003.* Needham, MA: Sloan Consortium.

Ally, M., & Coldeway, D.O. (1999). Establishing competencies and curricula for the distance education expert at the master's level. *Journal of Distance Education, 14*(1), 75-88.

American Distance Education Consortium (ADEC). (2004). ADEC guiding principles for distance learning. Retrieved January 5, 2005, from *www.adec.edu/admin/papers/distance-learning_principles.html*

Buford, J.A., Jr., & Lindner, J.R. (2002). *Human resource management in local government: Concepts and applications for students and practitioners.* Cincinnati, OH: Southwestern.

Dooley, K.E., & Lindner, J.R. (2002). Competency-based behavioral anchors as authentication tools to document distance education competencies. *Journal of Agricultural Education, 43*(1), 24-35.

Dooley, K.E., Lindner, J.R., & Richards, L.J. (2003). A comparison of distance education competencies delivered synchronously and asynchronously. *Journal of Agricultural Education, 44*(1), 84-94.

Drawbaugh, C.C. (1972). A framework for career education. *Journal of the American Association of Teacher Educators in Agriculture, 13*(2), 16-23.

Egan, T.M., & Akdere, M. (2004). Distance learning roles and competencies: Exploring similarities and differences between professional and student perspectives. In T.M. Egan, & M.L. Morris (Eds.), *Proceedings of the Academy of Human Resource Development 2004 Conference* (pp. 932–939). Bowling Green, OH: Academy of Human Resource Development.

Hanna, D.E. (1999). *Higher education in an era of digital competition: Choices and challenges.* Madison, WI: Atwood Publishing.

Lindner, J.R. (1999). Usage and impact of the Internet for Appalachian chambers of commerce. *Journal of Applied Communications, 83*(1), 42-52.

Lindner, J.R., Dooley, K.E., & Murphy, T.H. (2001). Discrepancies in competencies between doctoral students on-campus and at a distance. *American Journal of Distance Education, 15(2)*, 25-40.

Lindner, J.R., & Murphy, T.H. (2001). Student perceptions of WebCT in a Web-supported instructional environment: Distance education technologies for the classroom. *Journal of Applied Communications, 85,*(4), 36-47.

MacWilliams, B. (2000). Turkey's old fashioned distance education draws the largest student body on earth. *Chronicle of Higher Education, September 22*, A41-42.

McNeil, D.R. (1990). *Wiring the ivory tower: A round table on technology in higher education.* Washington, DC: Academy for Educational Development.

National Education Association (NEA). (2000). *Quality of the line: Benchmarks for success in Internet-based distance education.* Washington, DC: The Institute for Higher Education Policy.

Saba, F. (2003). Distance education theory, methodology, and epistemology: A pragmatic paradigm. In M.G. Moore & W.G. Anderson (Eds.), *The Handbook of Distance Education* (pp. 3-20). Mahwah, NJ: Lawrence Erlbaum Associates.

Simonson, M., Smaldino, S., Albright, M., & Zvacek, S., (2003). Teaching and learning at a distance: Foundations of distance education. Upper Saddle River, NJ: Merrill Prentice Hall.

Smith, P.C., & Kendall, L.N. (1963). Retranslation of expectations: An approach to the construction of unambiguous anchors for rating scales. *Journal of Applied Psychology, 47*, 149-155.

Texas Higher Education Coordinating Board (THECB). (1999). Principles of good practice for academic degree and certificate programs and credit courses offered electronically. Retrieved January 5, 2005, from *www.thecb.state.tx.us/reports/html/0206.htm*

Thach, E. (1994). *Perceptions of distance education experts regarding the roles, outputs, and competencies needed in the filed of distance education.* Unpublished doctoral dissertation, Texas A&M University, College Station.

Thach, E.C., & Murphy, K.L. (1995). Competencies for distance education professionals. *Educational Technology Research and Development, 43*(1), 57-79.

Webopedia. (2002). The difference between the Internet and the World Wide Web. Retrieved October 8, 2003, from *www.webopedia.com/DidYouKnow/Internet/2002/Web_vs_Internet.asp*

Western Cooperative for Educational Telecommunications. (2001). Best practices for electronically offered degree and certificate programs. Retrieved January 5, 2005, from *www.wiche.edu/telecom/*

Williams, P.E. (2000). *Defining distance education roles and competencies for higher education institutions: A computer-mediated Delphi study.* Unpublished doctoral dissertation, Texas A&M University, College Station.

Williams, P.E. (2003). Roles and competencies for distance education programs in higher education institutions. *The American Journal of Distance Education, 17*(1), 45-57.

Yeung, A., Woolcock, P., & Sullivan, J. (1996). Identifying and developing HR competencies for the future: Keys to sustaining the transformation of HR functions. *HR. Human Resource Planning, 19* (4), 48-58.

Chapter II

Critical Issues for Educators and Trainers:
Developing a Philosophy of Education

   Making Connections

In our last chapter, we explored the competencies and best practices needed to be successful in distance education. This chapter will continue to lay this foundation with a discussion of critical issues for educators and trainers. One of the first things to consider when creating or taking a new course at a distance is your own philosophy of education. Reviewing research studies can help educators, trainers, and learners understand the applications and practices that work in this setting. The concept of distance education as an innovation and the impact of technology in a global society are important as we consider the audience, access, and impact of distance education. Questions to guide your thoughts for this chapter are: What is the "no significant difference phenomenon" and how does research provide the theory to help guide the design, delivery, and evaluation of distance learning programs?

Introduction

Have you ever written a statement of educational philosophy? Sure, there are major schools of thought regarding philosophies: liberal, progressive, behaviorist, humanist, radical, and analytical (Elias & Merriam, 1980). But we are talking about your own personal philosophy. What do you believe about instructing and learning (in general)? These beliefs serve as the foundation for designing, delivering, and evaluating distance education courses and programs. We will return to this question in the application exercise at the end of the chapter.

We mentioned previously that being the instructional leader in a distance setting requires a unique bundle of competencies. For example, an instructor needs to know how to make the best use of the technologies available in order to personalize instruction and actively involve students in the learning experience. That is fundamentally our belief (philosophy). This belief is formed and strengthened by our research and the research of others.

Designing interactive components for instruction and feedback, ensuring that the audio/video components are working properly, and being comfortable with the technology that serves as the interface and connection between you and the distant learner are just a few of the skills needed for success in programs of distance education. For some, these knowledge, skills, and abilities (competencies) may be new. Consequently, distance education as a delivery system often may be perceived currently as being an innovation. Can it be? If so, why? A little background is in order. This background and perspective can also affect your philosophy.

Distance Education as an Innovation

Many universities and corporations are installing digital infrastructure to reach new audiences through distance education (Murphrey & Dooley, 2000). Specifically, continuing education, academic courses, and full degree programs are being developed to meet demand from individuals seeking nontraditional access. In the United States, 93 "cybercolleges," or accredited institutions offering credit-granting courses online in 1993, were listed in *Peterson's Guide*. In just seven years, there were 1,000 degree and certificate programs available from nearly 900 institutions (Peterson's Lifelong Learning Group, 2000). According to the International Data Corporation, the number of people in the United States taking at least one college course over the Internet tripled to approximately 2.2 million (Thornton, 1999). That type of growth continues.

Distance education may not appear to be an innovation to some, but for many educators, trainers, and learners, technology-mediated delivery systems are "new." Rogers defines an innovation as "an idea, practice, or object that is perceived as new by an individual or other unit of adoption" (2003, p. 12). Influencing the adoption and the diffusion of any new practice requires an understanding of change and change theory.

Multiple theorists focus on change theory. Bates (2000) and Hord, Rutherford, Huling-Austin, and Hall (1987) have considered specific contexts in business and education settings. Roger's *Diffusion of Innovation* (2003) has also been widely used to provide a framework to advance understanding of the rate of adoption of distance education based upon the characteristics or attributes of the innovation. So, briefly, what is an innovation? What are common characteristics of an innovation that have to be considered in attempting to speed up, slow down, or redirect the adoption of an innovation? How do these characteristics affect the diffusion of an innovation, in this case, programs of distance education? What process is involved in deciding to accept or to reject a program of distance education? Let's find out.

Diffusion is the process by which an innovation is communicated through communication channels over time (Rogers, 2003, p. 19). The *innovation-decision process* is the "process through which an individual (or other decision-making unit) passes from first knowledge of an innovation, to forming an attitude toward the innovation, to a decision to adopt or reject, to implementation of the new idea, and to confirmation of this decision" (Rogers, 2003, p. 20). The process can be influenced by prior conditions, characteristics of the decision-making unit, the perceived characteristics of the innovation, and communication channels.

Rogers (2003) discusses five attributes that impact the rate of adoption: (1) relative advantage, (2) compatibility, (3) complexity, (4) trialability, and (5) observability. *Relative advantage* is the degree to which an innovation is perceived as being better than a previous idea (Rogers, 2003, p. 219). Among the factors considered as relative advantage for distance education programs and processes are "things like incentives for faculty to participate or the ability to reach new audiences". The main function of an incentive is to increase the degree of relative advantage; however, if an improper incentive is chosen, long-term adoption may not result.

The second attribute, *compatibility*, "is the degree to which an innovation is perceived as consistent with the existing values, past experiences, and needs of potential adopters" (Rogers, 2003, p. 240).

The third attribute, *complexity*, "is the degree to which an innovation is perceived as relatively difficult to understand and use" (Rogers, 2003, p. 257). The nature of technology used and people's unfamiliarity with that technology could have a negative effect on adoption of distance education. The rate of adoption is slower with more complex innovations.

The fourth attribute, *trialability* (sometimes called "divisibility"), "is the degree to which an innovation may be experimented with on a limited basis. New ideas, that can be tried on the installment plan, are generally adopted more rapidly than innovations that are not divisible" (Rogers, 2003, p. 258). Fortunately, in most programs of distance education, trialability has not been a negative factor affecting adoption, as programs may be tried on a small scale, whether with number of students, number of sites, number of courses delivered, and so forth. Trialability can be assessed through evaluation studies as well.

The last attribute, *observability*, "is the degree to which the results of an innovation are visible to others" (Rogers, 2003, p. 258). The degree to which learners in a distance education program "spread the word" accurately to others and the degree to which others understand what they perceive distance education to be are possible effects of observability.

Distance education technologies continue to increase rapidly in power while decreasing in cost. Technologies such as the World Wide Web and multimedia have the potential to widen access to new learners, increase flexibility, and improve the quality of instruction by achieving higher levels of learning, such as analysis, synthesis, problem solving, and decision making (Bates, 2000). "The view of distance education as an innovation provides an important means for understanding the phenomena of distance education, particularly from the perspective of those upon whom its acceptance depends: the faculty" (Dillon & Walsh, 1992, p. 6). How people perceive and react to these technologies is far more important than the technical obstacles in influencing implementation and use, which may be reflected in the perceived characteristics or attributes of the innovation. How do you perceive and react to these technologies? Your perceptions and reactions will influence your philosophy of education as well.

During a symposium sponsored by the Pew Learning and Technology Program, a group of innovators in the field pointed out that individualization is the key to innovation. There is a need for distance education to be learner centered, combining high-quality, interactive learningware, asynchronous and synchronous conversations, and individualized mentoring (Twigg, 2001). Instruction should be modularized, self-paced, and delivered anywhere. A second component is improving the quality of student learning. This includes meeting the needs of diverse learners, when, where, and how they want to learn. The key goal is engagement in the learning process. A need to assess knowledge/skill level and learning style is an important step in designing instruction that is individualized and engaging. However, research has shown that learning styles are not predictors of learner achievement (Simonson, Smaldino, Albright, & Zvacek, 2003). The third component is increasing access to education and training. This includes both degree-offering institutions and lifelong learning programs. Finally, it is believed that the costs of teaching and learning can be reduced by using online instruction. This involves a reliance on centralized development of course materials, enabling quality control and disaggregated instructional roles (Twigg, 2001).

The bottom line is that learners in distance education environments have the potential to learn as much and as well as those taught using traditional methods (Schlosser & Anderson, 1994). But creating instructional content is not a trivial activity. It is not just applying telecommunications to a traditional lesson sequence. Simonson, Smaldino, Albright, and Zvacek noted that "the focus needs to be on creating optimal learning conditions for each individual" (2003, p. 144). These individuals may be located anywhere. Is creating such learning conditions embedded in your philosophy of education?

 Internet Connections
http://www.center.rpi.edu/PewSym/mono4.html

This link is to the complete PDF article by Carol Twigg entitled *Innovations in Online Learning: Moving Beyond No Significant Difference.* This project was funded by a grant from the Pew Charitable Trusts.

Globalization and Distance Education

With widening access to information technology, distance education and technology-mediated instruction is being adopted around the globe (Vrasidas

& Zembylas, 2003). Programs that were only available in certain regions of a state within the United States of America are now accessible in other states and countries and vice versa. This poses important instructional dilemmas in cross-cultural communication and delivery strategies. These dilemmas are intensified by the increasing "interconnectedness and use of technologies and processes from all around the world by both organizations and people" (Christiansen, 2000, p. 1). The complex nature and evolving concepts of globalization make it hard to define. However, one definition of globalization that addresses the interrelatedness and complexity of factors included in the term is that

> "Globalization is a rapidly increasing social, cultural, political, and economic process of awareness, though not necessarily acceptance, of a global consciousness and interdependence by which people make decisions about their life, their work, and their physical environment, decisions affected or influenced by expansion and interconnectedness of linkages throughout the whole world, not just the region or country in which they live and work, and decisions that over time collectively result in social, cultural, political, economic, and environmental conse-quences, both intended and unintended" (Christiansen, 2002, p. 1).

Although globalization is not easily defined, as can be seen above, it has been often described in terms of the world capitalist economy, the nation-state system, and the global information system. We would like to focus on the last component as it is central to the "expansion and interconnectedness of linkages" identified as a factor in the first, inclusive definition provided. Global information systems have not created a homogenous society that will replace local social systems. Yet global information systems have impacted our ability to communicate and share information worldwide across political, geographi-cal, and cultural boundaries.

Global television and news agencies have increased our ability to view current events. Likewise, fashion, music, and movies are diffused across various nations and cultures. Electronic data exchanges also allow transfer of electronic funds and purchasing of merchandise with a few mouse clicks and key strokes. The concept of an information society connected increasingly by international telecommunications can also greatly impact instructors and trainers.

Open and distance education is believed to be an engine of globalization because it serves to challenge the foundations of modern educational systems (Edwards & Usher, 2000). These challenges include a blurring of boundaries between formal and nonformal education, instructor and learner, classroom and home, print text and digital text, and education and entertainment (Edwards & Usher, 2000). People are looking for a community of learners through a new social network connected by information and communication technologies. This connection requires access to computers, modems, reliable phone lines, satellite systems, and Internet service providers.

Although educational information can be developed and distributed through telecommunications globally, these technologies are only accessible to small percentages of people, primarily located in North America and Northern Europe. The concept of the information superhighway is more like a toll road. In 1998, approximately 80% of the world's population lacked access to telecommunications (Eisenstein, 1998). Unfortunately, that percentage probably is still fairly accurate in 2004.

We believe that distance education has a part to play in empowering and educating the world. Whether working in a university setting, developing training for a multinational company, or preparing for a profession in distance education, the diffusion of distance education programs and practices as an innovation and the impact of emerging communication technologies on globalization should guide your decisions. We cannot work in isolation; as educators and trainers, we must collaborate and share ideas about best practices for the design, development, and delivery of educational content worldwide. We will come back to this principle in the last chapter as we explore future directions in distance education from the perspective of university graduate students, university faculty, extension and training specialists, and professionals working in international settings. Does your philosophy of education take globalization into account?

For the next section of this chapter, we introduce the concept of the "no significant difference phenomenon." Beliefs and knowledge about the effectiveness of instruction offered at a distance impact a philosophy of education and are another critical issue to be addressed when instructing and learning at a distance.

No Significant Difference Phenomenon

Anyone who has worked in the field of distance education has probably heard questions about differences between the quality of traditional, face-to-face instruction and that of instruction using distance delivery methods. Those questions are addressed in a book and on a Web site called the *No Significant Difference Phenomenon* (Russell, 1999). Over 400 research reports are summarized on technology used for teaching at a distance. The bottom line of this compilation is that it is not the media but the methods used that impacts instructional effectiveness. Prior studies show that the learning outcomes for online education are equal or superior to those of face-to-face instruction (Allen & Seaman, 2003).

Emerging technologies have shifted the focus, however, to making it easier to use a more learner-centered approach. Although the studies outlined by Russell focus on the media, future research should concentrate on the instruction itself, because it is truly a critical factor in determining learner achievement (Simonson et al., 2003).

 Internet Connections
http://www2.ncsu.edu/unity/lockers/users/f/felder/public/Papers/Resist.html

To learn more about learner-centered approaches to instruction, check out Navigating the Bumpy Road to Student-Centered Instruction by Richard Felder and Rebecca Brent available at this link.

Some researchers consider these studies flawed and of little value. For example, Phipps and Merisotis (1999) published *What's the Difference? A Review of Contemporary Research on the Effectiveness of Distance Learning in Higher Education* and pointed out that much research supporting the no significant difference phenomenon was based on weak research methods. However, it is time to move beyond media comparisons and truly examine meaningful learning dynamics. Meaningful learning dynamics may include studies on self-directedness, learner motivation, patterns of engagement, and specific media effects for teaching tactile and affective domains of learning.

To illustrate the importance of faculty acceptance for the diffusion and adoption of distance education as an innovation, we have provided below some findings from research on faculty perceptions at a Research 1 university. These also include reflections on current trends in an industry setting with an activity for Thought and Reflection at the conclusion of the section.

Internet Connections
<http://teleeducation.nb.ca/nosignificantdifference>

This site provides selected entries from the book *The No Significant Difference Phenomenon* (Russell, 1999). The purpose of this site is to provide access to recent comparative studies by year and media types, courtesy of TeleEducation NB.

An Example: Instructional Impacts of Technology

A survey was conducted on faculty at a Research 1 institution in order to provide a snapshot of how advances in telecommunications have impacted higher education (Dooley & Murphy, 2001). Faculty members were asked, "How has the use of distance education technologies impacted the teaching/ learning process during the past 5 years?" This question sought to determine the current uses/applications of distance education technologies and the faculty perceptions on teaching and learning at a distance. Uses and applications of distance education technology reported by the faculty included (1) Web-based tracking simulations, (2) PowerPoint presentations for in-class, videoconferences, and Web-based courses, (3) other multimedia, such as animation, (4) course Web pages with features such as the ability to check grades, download lecture outlines, class notes, handouts, course assignments, and course syllabi, and (5) e-mail and threaded discussion for increased communications and interaction between faculty and students. Faculty also mentioned using interactive video for guest speakers, including international connections, and using it for defenses of theses and dissertations in final graduate degree examinations.

Faculty perceptions varied. One professor noted, "Obviously, distance education technologies have had a tremendous impact on teaching in the past 5 years. Many courses are taught wholly via live video on the Web." Another stated that "It has been the biggest revelation in teaching in the past 100 years." Some faculty mentioned the added workload associated with teaching at a distance. "Courses have been offered and made available to students ... Yet, the instructor still shoulders much of the responsibility for coordinating equipment use at distant sites. Logistical support on campus and at distant locations is not adequate. Too many failures occur." Instructors also mentioned some limitations to teaching with technology: "Some courses, particularly with laboratories, are difficult to teach via electronic medium, because the students can't get firsthand experience in areas of learning that require feeling, touching, smelling, or similar sensations. Distance education is visual and auditory but must be combined with reading and analytical thinking for effective learning to occur." Another cited the common concern about lack of interaction or feelings of isolation between instructor and learner: "As a teacher, I wonder if the students have really learned or are going through the motions when I have less opportunity to get at shades of interest, understanding, doubts, and confusions in settings where I don't have strong and frequent face-to-face contact with students" (Dooley & Murphy, 2001, p. 7–8).

Trainers have similar concerns. Rightsizing, downsizing, reengineering of the company—all are terms that frighten today's employees. As companies are paring down their workforces, the discriminators among employees seem to boil down to a portfolio that includes education and training. Managers today must engage in some type of distance training if they are to narrow the educational and training gap of their employees. Corporations are now reevaluating the potential, and exploring the effectiveness, of a variety of instructional settings for human resource development. Distance education has become a strategic means for providing training, education, and new communications channels to businesses. A consideration of training at a distance forces a reexamination of the ways in which people learn and are trained. In the future, corporate employees will need to take control of their own growth and development, demanding training time and money as part of their rewards. Adult education principles, such as self-directed and lifelong learning, will become a major part of compensation packages (Dooley, Dooley, & Byrom, 1998). As this happens, increasingly, people will turn to or seek out programs of distance education.

Whether you are in a formal educational setting or a corporate training environment, the distance education industry is in a constant flux. Technologies will change, but the fundamental principles of effective instruction will remain the same.

Thought and Reflection

This example involves a privately owned general contractor with construction projects throughout the United States and in many foreign countries. Projects range in size from 50-800 employees, with most employees being unskilled and semi-skilled labor. Supervisors typically have minimal training and education. Supervisory training is a vital element in the corporate strategy to ensure that supervisors manage in accordance with corporate custom, regulations, and guidelines. Previously, this has occurred through on-the-job training (learn as you go), job-site training (make it up as you go), and corporate training (pay as you go). The lack of consistency and standardization is causing high employee turnover. Could distance education be a solution? If so, how?

Writing an Educational Philosophy

In order to implement good instructional principles and practices, the first step is to reflect upon and write down your own philosophy of education. Doing so will help you focus on why you do whatever it is that you do or will do in choosing and putting into practice sound instructional principles and practices. If you have never done this before, check out the Internet Connection below for a link to help you. If you have written one before, take it out, read it again, and see if it needs some revision as to what you currently believe and how it applies within the domain of instruction at a distance.

When writing the philosophy, think of one experience where you led or facilitated instruction that you would call your best. Why do you think this was the case? What approaches seemed to work well for you and the learners? Now, do the same thing with what you consider to be your most valuable learning experience. Describe the instructor–student relationship, the activities or products produced, how it was evaluated, the overall learning environment, and whether you were able to transfer this experience to other settings.

Internet Connections

http://www.utep.edu/~cetal/portfoli/writetps.htm

This link provides additional information about writing a personal philosophy of teaching from the University of Texas, El Paso.

Conclusion

In this chapter, we focused on aspects of philosophy of education and we reviewed the concept of distance education as an innovation. This concept provides the framework for other critical issues, such as the impact of distance education on globalization. Educators and trainers must be willing to participate in distance education and believe that distance education is as effective as traditional instruction.

Previously in this chapter, we asked "What is the 'no significant difference phenomenon' and how does research provide the theory to help guide the design, delivery, and evaluation of distance learning programs?" If you are a researcher in distance education, we hope that you will focus your efforts on creating optimal learning conditions for each individual (Simonson et al., 2003). If you are a practitioner or learner, we hope that you will take a look at some of the journals in the field such as the *American Journal of Distance Education*. Use the results of some of the research studies reported to guide your practice. Also consider how your own experiences may or may not be validated by research and how your own experiences have influenced your philosophy of education. You may wish to share these experiences in trade journals, professional meetings, or through training programs with others. Please see below a list of possible sources:

- *Innovate: Journal of Online Education (http://horizon.unc.edu/TS)*
- *International Journal of Technologies for the Advancement of Knowledge and Learning (www.TechKnowLogia.org)*
- *International Review of Research in Open and Distance Learning (www.irrodl.org)*
- *Syllabus Magazine (www.syllabus.com)*
- The Sloan Consortium *(www.sloan-c.org/index.asp)*

After a general discussion of learning theories, we will move into the realm of adult education. We hope that your philosophy of education espouses these values.

Application Exercise

Writing a philosophy of teaching is a valuable exercise. It requires reflective introspection. Jot down YOUR philosophy. It can be short—certainly no more than a page, and possibly a single sentence. It should be something you believe can be used as a touchstone in all the other decisions you will make in designing and delivering your distance education course or program.

Here is an example:
In order to prepare society-ready graduates and promote life-long learning, I strive to incorporate the latest innovations and telecommunication tools in my courses. Students work in teams, write and plan marketing strategies, give presentations using videoconferencing and multimedia, and basically wrap their knowledge/content acquisition with practical communication processes. In my classes, I want to instill the value of self-directed teams, empowered employees, and "boundaryless" organizations. I believe that learning is an active process. The learner is not just a mechanical processor of knowledge; s(he) also *interprets* the knowledge and information to build *meaning*.

References

Allen, I.E., & Seaman, J. (2003). *Sizing the opportunity.* Needham, MA: Sloan Consortium.

Bates, A.W. (2000). *Managing technological change.* San Francisco: Jossey-Bass.

Christiansen, J.E. (2000). *Globalization: What is it?* Departmental Leaflet 646-1-3b. College Station: Department of Agricultural Education. Texas A&M University.

Christiansen, J.E. (2002). *Globalization: A comprehensive definition.* Departmental Leaflet 646-1-3c. College Station: Department of Agricultural Education, Texas A&M University.

Dillon, C.L., & Walsh, S.J. (1992). Faculty: The neglected resource in distance education. *American Journal of Distance Education, 6*(3), 5-21.

Dooley, K.E., & Murphy, T.H. (2001). College of agriculture faculty perceptions of electronic technologies in teaching. *Journal of Agricultural Education, 42*(2), 1-10.

Dooley, L.M., Dooley, K.E., & Byrom, K. (1998). Unanticipated attitudinal change: The progression toward self-directed distance training at H.B. Zachry. In D. Schreiber, & Z. Berge (Eds.), *Distance training: How innovative organizations are using technology to maximize learning and meet business objectives* (pp. 351-368). San Francisco: Jossey-Bass.

Edwards, R., & Usher, R. (2000). *Globalisation and pedagogy: Space, place and identity.* London: Routledge.

Eisenstein, Z. (1998). *Global obscenities: Patriarchy, capitalism and the lure of cyberfantasy.* New York: New York University Press.

Elias, J.L., & Merriam, S. (1980). *Philosophical foundations of adult education.* Malabar, FL: Krieger Publishing.

Holmberg, B. (1995). The evolution of the character and practice of distance education. *Open Learning, 10*(2), 47-53.

Hord, S.M., Rutherford, W.L., Huling-Austin, L., & Hall, G.E. (1987). *Taking charge of change.* Austin, TX: Southwest Educational Development Laboratory.

Murphrey, T.P., & Dooley, K.E. (2000). Perceived strengths, weaknesses, opportunities, and threats impacting the diffusion of distance education technologies in a college of agriculture and life sciences. *Journal of Agricultural Education, 41*(4), 39-50.

Peterson's Lifelong Learning Group. (2000). *Peterson's guide to distance learning programs, 2000.* Lawrenceville, NJ: Thomson.

Phipps, R., & Merisotis, J. (1999). *What's the difference? A review of contemporary research on the effectiveness of distance learning in higher education.* Washington, DC: Institute for Higher Education Policy.

Rogers, E.M. (2003). *Diffusion of innovations* (5th ed.). New York: The Free Press.

Russell, T.L. (1999). *The no significant difference phenomenon.* Raleigh: North Carolina State University, Office of Instructional Telecommunications.

Schlosser, C.A., & Anderson, M.L. (1994). *Distance education: Review of the literature*. Ames: Iowa State University, Research Institute for Studies in Education.

Simonson, M., Smaldino, S., Albright, M., & Zvacek, S. (2003). *Teaching and learning at a distance: Foundations of distance education.* Upper Saddle River, NJ: Merrill Prentice Hall.

Thornton, C. (1999). Back to school, Web-style. *PC World, 17*(7), 39-40.

Twigg, C.A. (2001). *Innovations in online learning: Moving beyond no significant difference.* Troy, NY: The Pew Learning and Technology Program.

Vrasidas, C., & Zembylas, M. (2003). The nature of technology-mediated instruction in globalized distance education. *International Journal of Training and Development, 7*(4), 271-286.

Chapter III

Learning Theories

with
Tim Murphy, Texas A&M University, USA

 **Making Connections**

In the last chapter, we discussed critical issues that impact the design and delivery of distance education programs. You were asked to write your own philosophy of education statement to serve as the foundation for your instructional decision making. For our final chapter in Part I -Foundations, we will introduce the historical and philosophical frameworks that continue to guide and direct instructional decisions—learning theory. Consider these questions as you read: Which learning theory best matches my own philosophy of education? What are some assumptions about the nature of learning that are built upon theory? How can I apply learning theory when developing and using practices in distance education?

Introduction

Many learning theories guide our understanding of the learning process. While instruction and learning are explicitly correlated, the philosophy of learner-centered instruction is that educators and trainers should focus on the need of the learner. Gage (1972) noted that instruction focuses on methods used by educators to influence learners and that learning focuses on how learners learn over time.

To be effective in distance education, instructors should have examined their own beliefs about how people learn and how the learning process can be fostered in others. That is why we had you write your own philosophy of education in the application exercise at the end of chapter II. We suggest that both the distance education instructor and the distance education learner take time to think critically about how learning occurs and what they (individually and collectively) can do to facilitate learning. "Learning is a complex set of processes that vary according to the developmental level of the learner, the nature of the task, and the context in which the learning is to occur" (Gredler, 1997, p. 13). Models of learning introduced in this chapter include examples from behavioral, cognitive, and social learning theories.

Developing a Model of Learning: From Traditional Wisdom to Scientific Theory

How does learning occur? How can we explain, predict, and improve learning, both in ourselves and in others? To be an effective instructor, you must answer these questions for yourself. While perhaps not the earliest of all questions, people have been asking them for ages. They have answered them by developing models of learning from the sources of knowledge most accepted in their time. We will examine the foundations of each of the major sources of knowledge and some criticisms of the models of learning developed from each. The goal is for you to develop you own model of learning to guide you in your decision making about applications and practices in distance education.

Learning was modeled using folklore or traditional wisdom. General rules governing learning were passed down through a mostly oral tradition. These rules were applied to guide all learning activity. For instance, learning was

Thought and Reflection

HELPING PEOPLE LEARN

Educators can enable learning in a variety of ways.

- Train or teach when possible to a learner's strengths. If a learner demonstrates a propensity for science, develop tasks and projects to encourage and further their work. Encourage the successes of the learners while de-emphasizing the failings.
- Regular review of past lessons is often critical for concepts to be retained.
- Use approaches that draw on several disciplines at once to demonstrate how knowledge is built by casting wide for its foundation.
- Try to understand the neural networks that learners possess, realizing that incomplete networks just need additional inputs to make them whole. Thoughts and ideas can be found to be incomplete - just like sentences.
- Metaphors and analogies will often remove ambiguity from a concept. Draw on the life experiences one has had and do not hesitate to ask the class for their input from life.

Source: Zull (2002)

thought to require "discipline" and "hard work." The thing about "maxims" such as these is that there always is some truth in them, or they would not perpetuate themselves. The problem is they generalize across all settings and they lack specificity. This lack of specificity means that they are interpreted and applied in many different ways, some of which are inappropriate.

Models of learning based on traditional wisdom would suggest that the best source of information about any topic is someone who is good at it. Therefore, the best source about teaching would be a good teacher. On first encountering this, it may seem to be a good idea. What is wrong with this premise? It implies that the practice is perfect—that it never changes, that good teachers teach the same way, that resources do not have much effect. If you are working in the field of distance education, you know that the technologies and practices are constantly changing.

People eventually moved beyond models of learning based on traditional wisdom to models based on organized beliefs or organized myths. Over time, the organized belief systems that were logically consistent became philosophies. Models of learning were then proposed as part of the epistemology of each of the major philosophies. Each philosopher defined the nature of reality, and then answered questions such as "What is truth?" "What is knowledge?" and "What is learning?" They developed their answers in a manner consistent with their definitions of reality. The result was a logical and unified view of the external and internal worlds of the individual. Models of learning were then constructed based on these unified views. What could be wrong with this approach?

The information available in the philosophy to guide the development of a model of learning is normally limited because only general questions are asked or answered. While logically constructed from the philosophical worldview, the learning models are not tested in the real world, and therefore errors arise in the conclusions regarding cause–effect relationships. It may seem to be logically consistent that if you could learn your multiplication table by rote, you could also learn derivations in calculus using the same method. While no longer a mainstay of learning theory research, many of the ideas regarding the nature of knowledge identified by philosophers are included in modern theories of learning.

Early in the 20th century, empiricism, the complete reliance on research, the utter trust of truths found using the scientific method, swept through all fields, and education was not excluded. Models of learning were developed using science as the source of knowledge.

Models of learning were developed from research on animals (Thorndike, 1913) and children (Watson, 1925). Early scientific models of learning focused exclusively on establishing causal relationships between environmental stimuli and the behavior of the learners. These early researchers established models of learning that have since come to be called "behaviorist models." Later, other researchers, believing that the behaviorist models were too simplistic, became interested in the internal processes, or the cognitive activities, that learners go through. These researchers developed models of learning that came to be called "cognitive models."

Science has provided us with astounding advances in many fields; so what could be wrong with basing our models of learning on scientific evidence? Basing a predictive model on current research can produce results suggesting instructional practice as a patchwork of poorly fitted parts. Vygotsky (1993) suggests that the collection of data fails to advance knowledge about important phenomena. He attacks what he calls the "pedagogical anarchy" that results from efforts to utilize an "uncoordinated compendia of empirical data and techniques" to guide research and practice in learning and instruction (p. 93).

Another problem with using current research to guide your development of a model of learning is that there are often unknown or weak causal relationships between learning and the variables considered in educational research. We measure, for example, the number of books or computers per student in a school to assess the quality of the instructional environment. Would you like to try and predict learning from that data? Properly done, research contributes to knowledge about learning, but it is not a perfect platform for the development of models of learning or instructional practice.

So, while not appropriate by itself, research does enable and verify a source of knowledge that is considered appropriate for the development of predictive models of learning. The appropriate source of knowledge to use in developing your model of learning, and to guide your instructional practice, is theory grounded in research. "There is nothing so practical as a good theory" (Lewin, 1951, p. 169). Moreover, new theory does not emerge quickly but will be developed over time as research is extended and more and more data from learning are collected and analyzed (Dooley, 2002).

Theory can provide a framework for understanding and explaining research findings. While research attempts to explain the what is, theory attempts to explain the why. Theory can help distance educators understand why a particular instructional method is almost always effective in a particular setting.

As a source of knowledge, theories are superior specifically because they are based on research. Theories are therefore "testable." They can be supported (often called proven), and they can be discredited (disproven). Ned Flanders (1970) conducted research about the role of verbal interaction and learning. He measured amounts of various types of verbal interaction (talk) initiated by both instructors and learners. He found correlations to both learner achievement and learner satisfaction with specific kinds of talk. He concludes that learners' questions, and learners' talk on the subject, are very highly correlated with both achievement and satisfaction. Flanders' work supports the many theories of learning that contend that interaction between instructors and learners is a critical element. Research has extended this into distance education as one of the more critical elements in a distance education setting, whether it be learner-to-instructor, learner-to-content, learner-to-learner, or learner-to-technology. These interactions will be explained further in chapters IV and IX.

An essential element of a good theory is that it can be translated into concrete research designs, avoiding the random collection of data. Research is conducted in an orderly fashion to accomplish identified objectives to either support or discredit the theory or particular hypotheses underlying the theory.

A model developed on theory may accurately describe and predict learning today, but at some later date it may be seen as having value only in describing the historical development of the current model.

Theories are best used to develop your own working model of learning and instructing. As society changes and new information is discovered, theories are replaced, and the models based on them must be altered. Consider these questions: Why does learning a rule in one situation not guarantee that someone will be able to use it in another situation? Should I use "praise" in an online course? How should I organize my presentation of materials so that learners will retain more? Is this computer simulation an effective learning tool? How much interaction is required to facilitate learning? No single theory of learning adequately encompasses this variability. Each theory begins with some assumptions about the nature of learning and then develops a set of principles consistent with these assumptions. Your model of learning and instructing will almost necessarily encompass more than a single theory of learning.

What we hope you take away from this chapter is increased sensitivity to the need to evaluate research and methods of facilitating the learning process on sound theoretical bases and increased ability to use such evaluation to develop and modify your own model of instruction for learners at a distance. Other chapters will provide you with the necessary knowledge, skills, and abilities to see this model through.

In the next section of this chapter, we will examine three theoretical models of learning. Each of these models is still widely used in support of instructional methods. Each addresses learning in a different way, and in many cases, each suggests different instructional methods. We will progress through the three models in the order in which they were developed.

Behaviorism

Perhaps the most widely known of the behaviorist theorists is B.F. Skinner (1953, 1968). In Skinner's view, learning occurs when a response behavior is reinforced and thus the probability the response will occur in the future increases. Instruction is essentially the control of behavior, and occurs when the instructor reinforces the response he or she desires.

Building on the work of Thorndike, Skinner's work differs in one key component. All of these earlier theories suggest that "cues" or "drive stimuli" trigger behaviors from learners like some "inexorable force." In this view, desired behaviors can be "stamped in" through repeated practice. Skinner agreed. He thought this model accurately explained the learning of simple behaviors. Skinner then defined a new class of more complex human behaviors called "emitted responses." Examples of emitted responses include complex human behavior such as painting a picture or playing a song. He suggested that emitted responses act on the environment to produce different consequences. These differences in consequences alter future behavior. These behaviors that operate on the environment are called "operant" behaviors.

In Skinner's model, three components are essential in behavioral change. They are (1) the occasion on which the response occurs, (2) the subject's emitted response, and (3) the reinforcing consequences. In Skinner's model, it is the relationship between the emitted response and the reinforcing stimulus that alters the behavior, not the relationship between the stimulus and the response. These more complex operant behaviors are not linked "backward" toward some single stimulus, but rather "forward" toward the consequences they are expected to create in the environment. Skinner proposed that Thorndike had identified all the necessary components, but had misassigned the causal relationship for these more complex behaviors. In operant conditioning, behaviors are caused not by a stimulus but rather by the consequences they create in the environment.

Skinner also had much to say on the topic of reinforcement. To be effective in changing behavior, Skinner found that reinforcement must follow the desired behavior immediately. He also found that a consequence or reward was reinforcing if, and only if, it led to an increase in the frequency of the behavior rewarded. Many times we make assumptions about whether a reward will be reinforcing, and we are simply wrong. Whether a learner finds a particular reward reinforcing is a very individual matter. As an example, in our experience, many learners do not find grades reinforcing. Skinner is also widely misunderstood on the subject of negative reinforcement and punishment. These are not synonyms. Negative reinforcement is the removal of a consequence that the learner found rewarding, whereas punishment is the introduction of a stimulus the learner will find unpleasant.

Skinner suggested two procedures for facilitating learning through operant conditioning: shaping and chaining. Shaping is the reinforcement of successive approximations to a goal behavior. If you cannot get there from here, at least get closer. In other words, we provide reinforcement for a simple behavior that gets closer to the desired complex behavior. In teaching keyboarding, for example, an instructor may reinforce the learning by helping the student place his or her fingers on the correct keys to begin. Chaining is to string together individual simple behaviors that the learner already knows to establish new complex behaviors. An example of chaining is the design and layout of this book. The reader will advance through each topic with his or her lesson in mind and build on each with subsequent chapters until the end of the book when he or she will have a completed lesson ready for delivery at a distance.

Much of the current practice of instruction is based on behavioral learning theory. The entire field of instructional design is deeply immersed in stimulus, response, and consequences. Now we will examine another model of learning—cognitive information processing.

 Internet Connections
http://www.utm.edu/research/iep/b/behavior.htm

This link will take you to the Internet Encyclopedia of Philosophy for a detailed discussion of behaviorism. There are links to other philosophers and educators who shaped this model of learning listed with links to more information about this topic.

Cognitive Information Processing

As we look to the next model of learning, cognitive information processing, Skinner's model is useful in that, unlike his predecessors, he did not deny that "internal states" in the mind of the learner exist; he just said that they "lead us away from science" to "questions that should have never been asked" (Skinner, 1987, p. 782). In Figure 1, Skinner suggests that all cognitive learning theory fits inside the "black box." Black box metaphors are widely used in other fields. In biology many cellular process are "black box" processes. It simply means that no one knows what goes on inside, and we need not know to proceed. We do not know how a cell makes a particular protein; we simply know that when presented with the correct materials in the correct proportions, it does make the protein. Skinner holds that cognitive learning theory is like that. It may be interesting to know what goes on inside the mind of the learner, but it would have little effect on the practice of helping him or her learn.

Theories of learning continued to evolve. A growing number of researchers were beginning to see behaviorism as unable to describe complex learning

Figure 1. Skinner's black box metaphor

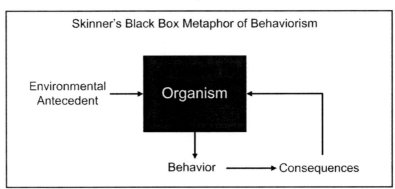

behaviors, and they were turning to cognitive models to attempt to overcome these perceived shortcomings in behavioral science theory.

In the 1960s, computers as large as train cars were being installed on college campuses and being applied to every sort of problem. Academics in every field began a love affair with numerical processing that has only grown in the ensuing four decades.

These new computers were almost instantly adopted by some learning theorists as "models" of the inner workings of the human mind. These learning theorists described their new field as cognitive information processing (CIP).

In most CIP models, including those espoused by Atkinson and Shiffrin (1968, 1971), learners process information in the same way a computer does. Information (or data) is "input" from the environment through the senses (or input systems), processed, stored in memory, and expressed (or output) in the form of behaviors. Figure 2 compares a basic computing model to the general learning model proposed by Atkinson and Shiffrin (1968). Each has five components. Input, short-term memory, processing, long-term memory, and output. In each, there are processes that are responsible for transferring information from one memory state or process to another.

In the CIP models, all human learning is derived from the environment. It is acquired through the five senses—visual, auditory, tactile, taste, and olfactory. We first see, hear, touch, taste, and smell everything in our environment that we come to know.

Although we sometimes talk about someone who has "magic fingers" on the piano, a "great eye" for identifying talent, a "great ear" for music, or a "green thumb" for tending plants, most of us would probably agree that these skills—

Figure 2. Comparing a mechanical computing model with the cognitive information processing (CIP) memory model in humans

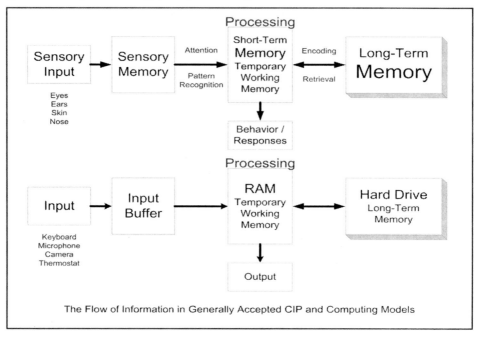

The Flow of Information in Generally Accepted CIP and Computing Models

all of them—actually do not reside in the fingers, the eyes, or the ears. They reside in the mind. The senses are simply (or really wonderfully complex) input devices.

Like most digital input devices, biological input devices briefly store information. The keyboard on a computer has a keyboard buffer. This is a small amount of digital memory that is used temporarily to store keyboard input until it can be processed. Digital video (DV) cameras utilize large amounts of temporary storage space when transferring DV to a computer's hard drive so that images and sounds will not be lost if there are unpredicted delays in the rate of processing or storage.

Our senses also use small amounts of memory for brief periods of time when capturing information for processing. Although still poorly understood, several experiments have been conducted to try and understand the role of sensory memory in CIP theories of learning.

Those working to advance CIP models espouse that all of our senses utilize some sensory memory and that these memories last for some brief amount of time. We are also fairly certain that the senses differ in both the amount of memory they possess and their duration. We also know that information presented using *more than one sense* is retained longer than information delivered using a single sensory channel. Clearly, in any model of learning, attention will be important. We simply cannot learn that to which we do not attend.

Attention is now believed to be a resource with *limited capacity* that is shared among all five senses, each of which can individually overload the capacity of our system. It appears that several ideas govern attention, the learner's ability to direct attention (often called "selective attention") and the ability of the mind to accomplish habitual complex tasks with little apparent attention (called "automaticity"). In all cases, no one processes all of the information available at any one time—we all miss something every minute. What we miss and what we process is a direct result of what gets our attention.

We all recognize that attention is selectively applied. We hear/see/feel much more information than we process. The sensory information that gets through to us is selectively chosen from all that is gathered. This process is partly under our conscious control, but it also operates automatically on predefined filters that we have established. If you have ever been intently reading a book in an airport, paying no attention to the noise of conversation and announcements around you, and then heard clearly the announcement that your flight will be delayed for another hour, you have experienced automatic selective attention. That you clearly hear your name, your hometown, or where you work announced in a far away conversation you had been ignoring is another example. Each of these items of information—your flight number, your name, and so forth—is preprogrammed into the search filters you have established to guide your attention. These filters run without conscious effort on your part.

The other process that governs attention is automaticity. This just means that some of us have done things, like drive a car or set up a computer, so often that we no longer need to use as much of our available attention to complete those tasks—in other words, they have become automatic. Remember if you can when you first started to drive. You were paying attention to the steering wheel, the road, the gas pedal, the speedometer, and so forth. Can you remember the first bridge you crossed while meeting another car? Remember how narrow it was? If someone (like your driving instructor) was talking to you at the same time, you never heard him or her. Now think about what we do after a few years of practicing driving. We talk to others in the car, we look in the back seat, we even use the phone! We adjust the radio, eat, put on makeup, read the paper. In other words, we pay attention to a great many things that are unrelated to driving a car. Is that because driving a car somehow takes less attention? No. It is just that the attention required to accomplish these driving activities has become automatic. If the driving activity changes, such as driving in a snow-storm, we have to assign new attention to these new activities.

Automaticity does not increase the total amount of attention available. The channel through which information passes from senses to short-term storage is not any wider. Automaticity allows us to off-load the attention required for habitual activity to a different pathway and frees us consciously to assign attention in other areas.

Clearly, attention must be secured before learning can occur. Good and Brophy (1984) recommend that instructors use standard signals, such as saying, "Let's begin," to alert students that they are to pay attention. While attention is required for learning in the CIP model, by itself it will not assure that environmental stimuli are available for processing in working memory. In order to facilitate this transfer to working memory, new information must fit a pattern already in memory. This is called "pattern recognition." Pattern recognition is the process in which environmental stimuli are recognized as examples of concepts and principles already in memory.

This is a preconceptual activity, meaning that there is no conscious thought involved. In order to process new information, the sensory memory has to recognize it. A match is made to something already in memory, but without identifying what either the new or the already-known object or idea actually is.

This phenomenon is not well understood, and several different models have been proposed to explain pattern recognition. We offer three of them, called "template," "prototype," and "feature recognition."

In template matching, it is thought that an exact mental copy of the stimulus is stored in memory. To recognize a person's face, you would locate an exact image of his or her face in memory. This really fails to explain our ability to recognize complex patterns or patterns that change. If someone cuts his or her hair, we can still usually recognize the person.

In the prototype model of pattern recognition, it is perceived that we store generalized prototypes instead of exact copies, and when new information matches all or most of a prototype, it is recognized. This explains how we might recognize someone who has cut his or her hair. He or she would still match most of our prototype.

The third model for pattern recognition is called "feature analysis." In feature analysis, we store only specific features or components of ideas or concepts and then search for these specific features in new stimuli. This model is gaining wide acceptance. In recent research on facial recognition, researchers have suggested that what we really store in memory is much more like a caricature of someone's face, emphasizing the features of the face that *make it different* than most faces. This model explains why we might recognize someone who cut his or her hair, and it also explains why we might *not* recognize him or her. If his or her hair was our *feature*, changing it would make him or her completely different.

When the senses pass information, when learners pay sufficient attention, and when an appropriate pattern is found in sensory memory, then new information enters the working memory. Here, for the first time, conscious processing can and does take place. At this stage, information is retrieved from the long-term memory and used in the processing of the new information. This process is primarily limited by the capacity of the short-term or working memory.

In 1968, Atkinson and Shiffrin described a multistore, multistage theory of memory. They suggested that information undergoes a series of transformations before it is stored in the long-term memory. This idea of "working" or "short-term" memory and "long-term" memory has been incorporated into almost every CIP model since. You can see the addition of this element in Figure 3.

Working memory has been described as a closet with a limited number of hangers. Nearly everyone (about 68%) has seven hangers in their closet. Miller

Figure 3. Cognitive information processing (CIP) model to explain acquisition of memory

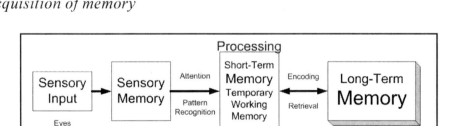

(1956) lightheartedly suggested that it was no accident that there are seven digits in a telephone number. He also wondered if people could remember more, would there be only seven wonders of the world? Seven primary colors? Seven seas?

All kidding aside, seven bits of information does seem to be an important number in describing the limits of working memory for the average person, and not just for meaningless strings of numbers. Other kinds of information—letters, images, sounds—all respond to Miller's "seven plus or minus two" limitations.

Clearly, we would be hard pressed to learn anything very complex if we could only store seven discrete numbers or letters at one time. So why is it that we are able to process much larger pieces of information? Miller found that while there are only seven hangers in your closet, *what you hang on each of them is almost unlimited.* For instance, look at these fifteen letters USAIBMNASAHP. Difficult to remember? How about USA, IBM, NASA, HP? As four chunks, the twelve letters are easily remembered. Miller found that working memory capacity may be increased through creating larger chunks. This process has been called "chunking," and is critical to many models of instructional design.

As anyone who has even been given a Web site address or a phone number knows, working memory does not last forever. In other studies, the limit was found to be 15–30 seconds before all information was lost (Peterson & Peterson, 1959). If we want to remember something we have just been told, such as a Web site address, we need to do two things: rehearse and encode.

Rehearsal is just what it sounds like. Repeat the Web site address five times. While this certainly helps us remember an address for short periods, to move information into long-term memory we need to "encode" it. Encoding is the process of relating new information to concepts and ideas already in memory in a way that makes it more memorable. There are many ways to help learners encode information. Any organization at all, even one that seems nonsensical, has been shown to improve learning. How many of you can name the colors of a rainbow in the order they appear? (ROY G BIV). These mnemonic devices are nonsensical, but they serve to help remember information that is not. Information arranged according to some logical organization, such as time (chronology), importance (hierarchies), size, and so forth, helps learners remember even more for even longer periods of time. Mental images have also been shown to aid in remembering information. Simple, even seemingly silly, images help learners remember the ideas associated with them. Think about the signs for radiation and for Mercedes-Benz. Similar shapes but vastly different meanings. Diagrams or concept maps of ideas aid learners in remembering the ideas associated with the diagram.

While senseless mnemonics and diagrams will help, the best way to help learners encode is to provide meaningful connections to information that the learners already have. Comparing and contrasting new ideas to old will encourage encoding. Ask questions such as "How is this different than ... ?" or "The components of this idea are very similar to the idea we discussed last week ..." Through effective rehearsal and encoding, information is moved into long-term memory. One thing to keep in mind about long-term memory is that it has no known limit.

 Internet Connections
http://library.thinkquest.org/~C0110291/basic/brain/short.php

This link will take you to a discussion about short term or working memory from *Think Quest*. There are also a few activities and exercises if you would like to try them.

You may not be able to recall what the weather was like last Tuesday, but we bet you can recall a lot about the weather on your wedding day or on the last day of a vacation taken years ago. Tulving (1972) makes a distinction between episodic and semantic memory. Memories that are learned in an environment, and cannot be recalled without also recalling the episode in which they were created, are episodic. All the memories of general information that can be recalled independently of how they were learned are semantic.

Several models of how the long-term memory actually stores information have emerged. This is an area of CIP theory that is still under development, and is fertile ground for additional research. The same is true of the models for retrieving information from long-term memory. There has been some interesting research in retrieval that can help explain some of the principles of creating test questions at various levels of difficulty, and in the differences in effectiveness of various study habits.

What does the CIP model say to the instructor? First, provide *organized* instruction. While several techniques are appropriate for this, Beissner, Jonassen, and Grabowski (1994) suggest that the use of instructional graphics is especially effective. They conclude that well-developed instructional graphics analyze, elaborate, and integrate subject matter, as well as illustrate relationships between concepts. So, use graphics such as cognitive maps, diagrams, and visual hierarchies to help learners process and encode.

Instructors then need to arrange for extensive and variable practice. While "practice makes perfect" is grounded in classical conditioning, specifically in the work of Thorndike, it applies here as well. As basic skills achieve automaticity, then conscious attention may be directed at more complex learning tasks. This is a worthwhile goal, and one way to achieve it is through extensive practice of the basic skills. It is appropriate that the words "Learning to do, doing to learn" are the first two lines of the four-line motto of the agricultural education FFA (Future Farmers of America) organization.

Finally, the CIP model suggests that in practice, we should help learners develop "executive control processes" or ways to improve their own learning. People need to notice, think about, and experiment with how they learn. They need to learn how they learn. Instructors can help by presenting strategies that have worked for others, and by simply helping learners explore and evaluate their own strategies.

In concluding this section on cognitive information processing (CIP), think about this. It is at least somewhat ironic that computers were adopted by some learning theorists as "models" of the inner workings of the human mind when the computer scientists working to design these new devices and the software that made them work began by asking themselves "How does the human mind work? How does biological intelligence work? And then they tried to emulate that as closely as possible in an electromechanical device. In fact, computer scientists today continue to mimic biological intelligence in silicon, carbon, and gold. The latest concepts in computer science, object-oriented software, distributed networking, and quantum-bit computing, all borrow their foundational ideas from biology. Observant outsiders might conclude that research in human learning has contributed to computer science as much as computer science has contributed to human learning.

The CIP model was built from or with the model of binary computing in vogue throughout the 1950s through the 1990s. We are on the brink of a whole new way to compute. Researchers have a working model of a quantum-bit computer in operation. These machines have three states, whereas a digital bit is either 1 or 0, a quantum bit can be neutral. Yes? No? Maybe? In theory, these machines will be thousands of times faster at complex data analysis and prediction (such as data mining and simulation). If you think about it, is it not human to think maybe? That is how our thinking differs from a machine. From maybe comes creativity. From maybe comes ingenuity. The CIP models attempt to fit human learning into yes and no to match a metaphor of digital computing that may already be nearly extinct.

 Internet Connections
http://chiron.valdosta.edu/whuitt/col/cogsys/infoproc.html

This Internet Connection will take you to Educational Psychology Interactive: The Information Processing Approach developed by W. Huitt. Cognitive Information Processing is discussed with links to additional references and resources. A discussion of memory is also included.

Situated Cognition

Situated cognition is grounded in the social learning theories. Social learning theory suggests "that people learn from observing other people. By definition, such observations take place in a social setting" (Merriam & Caffarella, 1991, p. 134).

Behaviorists looked at how people learned through observation. Later researchers, for example, Albert Bandura (1977), looked at interaction between and among learners and how that facilitated cognitive processes. Learning from observation allows people to see the consequences of others' behaviors. Imagine if people had to rely solely on the effects of their own actions to inform them what to do. What if we could only learn though personal discovery? Bandura (1977) stated that "it is fortunate that most human behavior is learned observationally through modeling: from observing others one forms an idea of how new behaviors are performed, and on later occasions this coded information serves as a guide for action" (p. 22).

Situated learning, as described by Lave and Wenger (1991), is one of the more radical social learning models. Rather than looking to learning as the acquisition of certain forms of knowledge, they argue that learning is *in* social relationships—situations of coparticipation with others. "Rather than asking what kind of cognitive processes and conceptual structures are involved, they ask what kinds of social engagements provide the proper context for learning to take place" (p. 14). It is not so much that learners acquire structures or models to understand the world, but that they participate in frameworks that have structure and they contribute to that structure.

According to this theory, learning involves participation in a community of others who are practicing the behaviors to be learned. Learning is situated in communities of practice and cannot be analyzed in isolation from either the practice or the community. There is an interconnectedness of persons, learning, practice, participation, and the social world.

Learning is then operationally defined as increased participation in these communities of practice. According to Lave, this definition "focuses attention on ways in which it [learning] is an evolving, continuously renewed set of relations ... [among] persons, their actions, and the world" (Lave & Wenger, 1991, p. 50).

Learning is a process affecting both the learner and the community in which he or she participates. Both are transformed by their interactions in the world. We have communities of practice around our work, our hobbies and social activities, every situation in which we participate with others. These communities of practice are themselves situated inside larger social organizations.

Lave and Wenger's (1991) examples of communities of practice are an example of groups that have a feeling of community due to a common *purpose*. They argue that communities of practice must share a joint enterprise. Members will probably also share some common background or experience and will certainly share a common language. As new members join the group, they learn from the existing members as they work. In such groups, the informal lines of communication have been shown to be critical for learning to take place and for new knowledge to be created. Within these informal networks, members will swap experiences and anecdotes and learn from each other.

Learning is not seen as the acquisition of knowledge by individuals so much as a process of *social* participation. The nature of the *situation* impacts signifi-

 Internet Connections
http://tip.psychology.org/lave.html

J. Lave provides a nice overview about situated learning or cognition. There are additional web links at the bottom of the page including the Cognition and Technology Group at Vanderbilt University.

Figure 4. Community of practice as a process of social participation and, hence, of learning

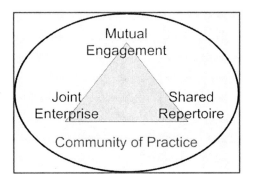

cantly on the process. Situated learning, then, is usually unintentional, rather than deliberate. Learners acquire mastery of the subject through "legitimate peripheral participation" in these communities of practice (Figure 4).

Learners inevitably participate in communities of practitioners and … the mastery of knowledge and skill requires newcomers to move toward more full participation in the sociocultural practices of a community. "Legitimate peripheral participation" provides a way to speak about the relations between newcomers and old-timers, and about activities, identities, artifacts, and communities of knowledge and practice. A person's intentions to learn are engaged and the meaning of learning is configured through the process of becoming a full participant in a sociocultural practice. This social process, includes, indeed it subsumes, the learning of knowledgeable skills (Lave & Wenger, 1991, p. 29).

The belief of situated cognition and social learning theory is that individuals learn through participating in communities of practice. Organizations and institutions learn through establishing and fostering these communities. Communities learn through refining practice and ensuring new members enter the community (Figure 5). Learning at the organizational level "is an issue of sustaining the interconnected communities of practice through which an organization knows what it knows and thus becomes effective and valuable as an organization" (Wenger, 1998, p. 8).

Figure 5. Individually lived practices for community learning

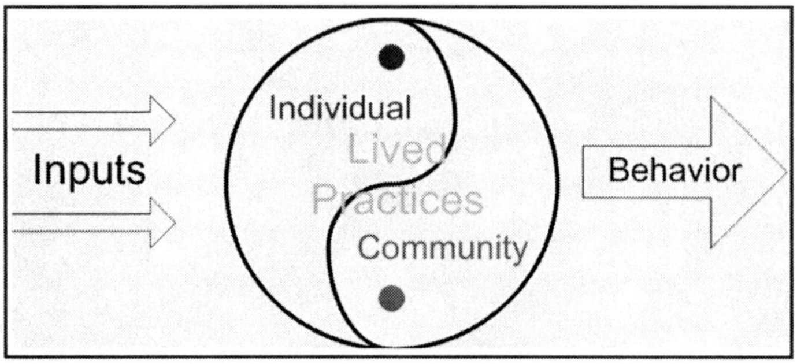

Implications of Learning Theory
for Educators and Trainers in
Distance Education

Now that we have shared three models of learning—behaviorism, cognitive information processing, and situated cognition—we will summarize some of the implications for educators and trainers in distance education settings.

The implications for behaviorism in designing and delivering education and training at a distance are multidimensional. On the one hand, teaching some very complex tasks that require the exact and same response every time, is classic behavioral theory. Examples would include teaching one to program in HTML or ASP code. There is one and only one way to write a command. On the other hand, shaping and chaining are ways to use a behaviorist approach and still allow the learner different avenues to arrive at the same answer. Also, as noted, the entire field of instructional design is heavily embedded in behaviorism. To observe the steps in instructional design, you have to recognize clearly stimulus, response, and consequences.

In using theory arising from cognitive information processing, it is important to recognize the differences in how the brain works and how individuals process information. How learners process short-term and long-term memory will affect how educators present material for learning. Should the instructor group similar topics or similar group experiences? What this says to the instructor is to provide a well-organized lesson. Moreover, the use of graphics, mind maps, diagrams, and so forth, will help learners to process and encode the material in a much more efficient manner.

Finally, in situated cognition we have been reminded that learning is a social process laden with social relationships. Instructors must recognize the interconnectedness of the learner, the learning, and the social world. It is important to design into your instruction these learning communities. An example would be team assignments where the learner will be a part of a team, but also could be "fired" from the team for not following the community rules.

Thought and Reflection

QUESTIONS INSTRUCTORS SHOULD CONSIDER

- How can we use our knowledge of the workings of the brain better to frame lesson plans and instruction?
- As instructors, shouldn't we use an integrated approach with assignments stimulating the brains of our learners by calling on their experiences and memories through reflection?
- How can we have our learners demonstrate their ideas in distance education environments?
- Active participation in the learning process is one way to acquire education. Why not strive to observe the learners actively testing their own ideas? Wouldn't this help them become involved in their own education?
- With active involvement, shouldn't we actively encourage learners to use all parts of the learning cycle?

Source: Zull, 2002

Conclusion

After providing a philosophical discussion on the development of models of learning, this chapter summarized three models: behaviorism, cognitive information processing, and situated cognition/social learning theory. There are many more theories available, so we encourage you to explore these concepts further using the Internet Connections or your own resources and Web searches. Fundamentally, examining and assessing learning theory should help you to better understand learner differences and characteristics that impact the effectiveness of instruction and learning. In Section II of this book, we will explore some of these differences as we begin to focus on the principles of adult education.

 Application Exercise

SHOULD WE HAVE A THEORY ON INSTRUCTING AND LEARNING FOR DISTANCE EDUCATION?

Does distance education require its own set of theories on teaching and learning? If yes, how should these theories be developed? What would be the philosophical underpinnings of such theory? If not, from where will the theories be borrowed and in what form should they be used? If borrowed, whose discipline will be chosen as donor? Which aspects of distance education need to be addressed first? Given the solitary nature of distance education, should teamwork be a part of a distance education curriculum? How can it be learned? (Moore & Anderson, 2003)

Take some time to consider the three learning theories posed in this chapter or search for others. Look back at your philosophy of education. Is there a match? You may want to add to your philosophy at this point to include a description of the model of learning that will guide you.

Begin to think about a content area that you would like to create or convert to distance education. Jot down the potential audience and topic you would like design. We will continue to add to this in future application exercises.

References

Atkinson, R.C., & Shiffrin, R.M. (1968). Human memory: A proposed system and its control processes. In K.W. Spence & J.T. Spence (Eds.), *The Psychology of Learning and Motivation: Advances in Research and Theory* (Vol. 2, pp. 89-195). New York: Academic Press.

Atkinson, R.C., & Shiffrin, R.M. (1971). The control of short-term memory. *Scientific American, 225,* 82-90.

Bandura, A. (1977). *Social learning theory.* New York: Prentice Hall.

Beissner, K.L., Jonassen, D.H., & Grabowski, B.L. (1994). Graphic techniques for conveying structural knowledge. *Performance Improvement Quarterly, 7*(4), 20-38.

Dooley, L.M. (2002). Case study research and theory building. *Advances in Developing Human Resources, 4*(3), 335-354.

Flanders, N.A. (1970). *Analyzing teaching behavior.* Reading, MA: Addison-Wesley.

Gage, N.L. (1972). *Teacher effectiveness and teacher education.* Palo Alto, CA: Pacific Books.

Good, T.L., & Brophy, J.E. (1984). *Looking in classrooms.* New York: Harper & Row.

Gredler, M. (1997). *Learning and instruction: Theory into practice* (3rd ed.). Upper Saddle River, NJ: Prentice-Hall.

Lave, J., & Wenger, E. (1991). *Situated learning: Legitimate peripheral participation.* Cambridge, UK: Cambridge University Press.

Lewin, K. (1951). *Field theory in social science; Selected theoretical papers* (D. Cartwright, Ed.). New York: Harper & Row.

Merriam, S., & Caffarella, R. (1991). *Learning in adulthood. A comprehensive guide.* San Francisco: Jossey-Bass.

Miller, G.A. (1956). The magic number, seven, plus or minus two: Some limits on our capacity for processing information. *Psychological Review, 63,* 81-97.

Moore, M., & Anderson, W.G. (Eds.). (2003). *Handbook of distance education.* Mahwah, NJ: Lawrence Erlbaum.

Peterson, L.R., & Peterson, M.J. (1959). Short-term retention of individual verbal items. *Journal of Experimental Psychology, 58,* 193-198.

Skinner, B. (1968). *The technology of teaching*. New York: Appleton-Century-Crofts.

Skinner, B.F. (1953). *Science and human behavior*. New York: Macmillan.

Skinner, B.F. (1987). Whatever happened to psychology as the science of behavior? *American Psychologist, 42*, 780-786.

Thorndike, E. (1913). *Educational psychology*. New York: Teachers College, Columbia University.

Tulving, E. (1972). Episodic and semantic memory. In E. Tulving, & W. Donaldson (Eds.), *Organization of memory* (pp. 381-403). New York: Academic Press.

Vygotsky, L. (1993). *The collected woks of L.S. Vygotsky: Vol. 2. The fundamentals of defectology (abnormal psychology and learning disabilities)* (J.E. Knox & C.B. Stevens, Trans.). New York: Plenum.

Watson, J.B. (1925). *Behaviorism*. New York: Norton.

Wenger, E. (1998). *Communities of practice. Learning, meaning and identity*. Cambridge, UK: Cambridge University Press.

Zull, J.E. (2002). *The art of changing the brain: Enriching the practice of teaching by exploring the biology of learning*. Sterling, VA: Stylus.

Section II

Adult Learning Theory

Section II of the text focuses on the ability of adults to be self-directed, a fundamental concept of distance learning. Incorporating adult-learning principles (andragogy) into the design and delivery of distance courses will result in more meaningful learning. Taking into account a learner's unique background, experiences, knowledge, skills, abilities, self-directedness, and/or personal styles and values will help ensure meaningful learning experiences.

Chapter IV

Adult Learning Principles and Learner Differences

with
Susan Wilson, Texas A&M University, USA

 **Making Connections**

A clear ideology for instructing and learning at a distance does not exist. An emerging belief by researchers and practitioners is that the use of andragogical principles and practices results in deeper and more meaningful learning by adults. We concur. As discussed in previous and subsequent chapters, how materials are delivered does not have an effect on learner achievement, but what methods are used to engage learners does. In chapter III, we explored models of learning and their application in distance education. In this chapter, we will introduce theory and practice that support the use of adult learning principles when instructing at a distance. We will also address strengths and weaknesses of andragogical and pedagogical methods. Questions to guide you in this reading include "What is the role of an educator when instructing at a distance?" and "How can educators foster deeper and more meaningful learning?"

Introduction

Educators and trainers should attempt to design and deliver individualized instructional sequences to provide the greatest opportunity for the learner. To achieve this lofty goal, educators and trainers will have to teach, coach, mentor, facilitate, motivate, and direct learners based on the educators' assessment of learners' unique backgrounds, experiences, knowledge, skill, abilities, personality type, social style, and/or personal styles and values (Lindner, Dooley, & Williams, 2003). Knowles (1990) suggested that as a person matures and ages, his or her dependence on an educator to teach decreases. The preceding statements are, in fact, essential for effecting instruction both in the classroom and at a distance.

While learner achievement may not be affected by how curricular materials are delivered, how learners interact among themselves and with the instructor does have an effect on learner satisfaction. Learner satisfaction improves as interactions among themselves and with the instructor increases (Fulford & Zhang, 1993; Garrison, 1990; Ritchie & Newby, 1989). While distance education may help instructors reach learners separated by location and/or time, *transactional* distance may hinder learner satisfaction and achievement. The concept of transactional distance was first discussed in 1980 (Moore) and continues to be a major barrier for the adoption and diffusion of distance education. Transactional distance is a measure of distance not as a geographical but as a "pedagogical phenomenon" (Moore & Kearsley, 1996, p. 200). It involves the interplay among the instructors, the learners, the content, and the learning environment. Distance is described in terms of the responsiveness of an educational program to the learner, rather than in terms of the separation of the instructor and the learner in space or time or both.

Distance education as a contextual application, we would argue, is mature. The widespread appeal and acceptance of online learning, however, has not resulted in changes necessary to maximize its effectiveness and efficiency (Howard, Schenk, & Discenza, 2004). Howard, Schenk, and Discenza further suggest that the majority of distance education courses use pedagogies developed for traditional face-to-face classes. "By clinging to traditional pedagogies, universities often diminish the potential education advantages brought by the technologies used for distance education" (p. vi).

Adult Learning Principles

Because distance education draws on the ability of learners to be self-directed, incorporating adult learning principles (andragogy) in the design and delivery of content results in more effective learning (Richards, Dooley, & Lindner, 2004). Andragogy is based on the following six assumptions about the learner: (1) learner's need to know, (2) self-concept of the learner, (3) prior experience of the learner, (4) readiness to learn, (5) orientation to learning, and (6) motivation to learn (Knowles, Holton, & Swanson, 1998). The box for Thought and Reflection provides an explanation of each of these assumptions.

Pedagogy, in contrast, relies on different assumptions about the learner and can be defined as the "art and science of teaching children" (Knowles, Holton, & Swanson, 1998, p. 62). Knowles, Holton, and Swanson noted that within the pedagogical model

> learners only need to know that they must learn what the teacher teaches if they want to pass and get promoted; they do not need to know how what they learn will apply to their lives. The teacher's concept of the learner is that of a dependent personality; therefore, the learner's self-concept eventually becomes that of a dependent personality.

While the distinction between the literal definitions of andragogy and pedagogy provides for interesting debates by academicians and researchers, it is, in fact, the distinction among the underlying assumptions that is important to understand. Educators who put their interests and needs (intentional or unintentional) over those of the learners, restrict meaningful learning. The ultimate goal of an educator should be to facilitate learning (Leamnson, 1999). This will require the educator to be a teacher, coach, mentor, facilitator, motivator, and/or authoritarian, depending on the learners' personal characteristics (Lindner, Dooley, & Williams, 2003).

 Internet Connections
http://agelesslearner.com/intros/andragogy.html

This site provides an excellent discussion about the similarities and differences between andragogy and pedagogy.

Thought and Reflection

KEY ASSUMPTIONS ABOUT THE LEARNER

Learner's Need to Know. Learners need to know why they need to learn anything before beginning the instructing and learning process. An educator's task is to help learners identify gaps between what they know and what they need to know. Self-assessment measures are particularly useful at the beginning of the learning process to help learners identify gaps in their competence.

Self-Concept of the Learner. Learners are ultimately responsible for the decisions they make regarding what they will learn. They do not want to be dependent on educators to "teach" them. Courses should focus on the self-directed nature of learners.

Prior Experience of the Learner. A learner's prior experience cannot be sequestered from the learning process. They will use their life experiences and common sense to facilitate learning. When possible, focus on educational activities that allow learners to draw on and share their prior experiences.

Readiness to Learn. Learning occurs best when the information to be learned can be directly applied to solving a problem or filling an identified gap. As much as possible, focus on information that can be used practically.

Orientation to Learning. Learners learn best when the educational process occurs in the context of real-life situations. Learners are task-oriented in their approach to learning; they want to be able to use what they learn to solve problems and accomplish tasks. Educators should be learner-centered, not topic-centered.

Motivation to Learn. Learners want to increase their competencies. They are motivated by internal motivators, such as learning for the sake of learning, self-esteem, enjoyment, or quality of life. To a lesser degree, they are also motivated by external motivators, such as higher pay, better jobs, or advancement opportunities.

Source: Knowles, Holton, & Swanson, 1998; Richards, Dooley, & Lindner, 2004

Applying Adult Learning Assumptions to the Educational Process

Educators can help learners identify gaps in their competence by using pretests. For example, we use a self-assessment instrument at the beginning of a class titled "Advanced Methods of Distance Education" to help learners identify where they believe their strengths and weaknesses are with respect to each of the core competencies covered in the course.

The use of learning contracts can facilitate the self-concept of the learner. Learning contracts are written agreements between an educator and a learner. The contract essentially describes what the learner will do and how that activity will be evaluated by the educator. Questions addressed in a learning contract include the following (Knowles, 1995): What are you going to learn? How are you going to learn? What is your target date for completion? How are you going to know that you learned it? and "How are you going to prove that you learned it?"

Developing learning activities that require learners to draw on and share their prior experiences facilitates deeper and more meaningful learning. Noted philosopher, educator, and author John Dewey (1938) stated that the education process begins with experience and that "all genuine education comes about through experience" (p. 13). For example, learner-led threaded discussion groups can be used to help learners think about how course materials can be used in various contexts.

The assumption about readiness to learn suggests that learning is situational. Learners often exhibit different behaviors in different learning situations (Pratt, 1988). Learners are more likely to blame their own poor performance on situational factors rather than their own deficiencies (Sabini, Siepmann, & Stein, 2001). Subsequently, people tend to overestimate the role of other people in producing behavior and underestimate the role of the situation. This tendency is referred to as the "fundamental attribution error." A learner's readiness to learn can be induced by providing learners with diagnostic experiences to evaluate how curricular materials can be applied to their lives.

Using learner-centered approaches facilitates a deeper and more meaningful learning experience than knowledge- or content-centered approaches (McKeachie, 1999). Using learning activities that require learners to solve problems that are meaningful to them is important. One way to do this, for example, would be to have learners use the targeting outcomes of programs

(TOP) model to evaluate an activity on which they are working outside of the learning environment (Bennett & Rockwell, 1995).

Thought and Reflection

Andragogy—An Alternate Interpretation
The concept of andragogy can be interpreted in several ways. To some it is an empirical descriptor of adult learning styles, to others it is a conceptual anchor from which a set of appropriately "adult" teaching behaviors can be derived, and to still others it serves as an exhortatory, prescriptive rallying cry. This last group seeks to combat what it sees as the use with adult learners of overly didactic modes of teaching and program planning, such as those commonly found in school-based, child education. Andragogy is also now, for many educators and trainers of adults, a badge of identity (Brookfield, 1986, p. 90).

Stimulating the learner's desire to participate in the learning process is critical. While external factors, such as grades, do motivate learners to learn, internal factors such as curiosity have a greater positive learning impact. Creating a learning environment that values and respects individual competence and experience is critical to motivating learners. Early in the learning process it is important to provide positive and frequent feedback to the learners as well as capitalizing on teachable moments.

Internet Connections
http://www-distance.syr.edu/contract.html

This site provides an excellent discussion about the use of learning contracts.

Learners' Differences

Are you ready to read about learners' differences and their importance or unimportance to distance education? If not, close your textbook and close your eyes.

Are you still reading? Why? Why are you still reading and your counterpart is sleeping and starting to snore? How are you different from your conked-out counterpart? Which difference matters most to you and your learning?

Overall, educators have agreed that those who learn differ in how they learn. Here, we will review research of learners' differences—in demography, learning styles, cognitive styles, cognitive controls, and intelligences—and answer that same question: "Which difference matters most to you and your learning, whether at a distance or in a classroom?"

Jonassen and Grabowski (1993) linked learners' differences, learning styles, and cognitive styles and controls in one simple sentence: "Student learning differs because student learning traits differ (sic), and because the thinking process differs depending on what the student is trying to learn" (p. 3). They listed learners' likely differences, to which we added, subtracted, and swapped differences to suit our own steps of deduction (Figure 1).

Differences in Demography: Age and Generation

Today's up-and-coming college learners, born in or later than 1982 and called "Millennials," share several aptitudes for and attitudes about technologies that are tied to education. First, Millennials like to learn hands on and by trial and error, just like they might try trial and error to solve problems by their Nintendos and Sony PlayStations. Second, Millennials multitask by taking on two, three, or more activities or assignments at once, for which they call for constant and continuous connections to their classmates, families, friends, and instructors. Third, Millennials combine and confuse creators and consumers with their cut-and-paste plagiarism and peer-to-peer piracy (Oblinger, 2003). Here, we will review research about the generation gap in abilities and education and relate that research to Baby Boomers, Gen-Xers, and Millennials.

In addition, Oblinger (2003) points out that more and more learners at two- and four-year colleges are at or older than 25, tagging them Gen-Xers and Baby Boomers. Three fourths of college learners are nontraditional, caring for families, waiting to come to college, and/or working full time. Of course, Baby Boomers and Gen-Xers likely look at education and technologies differently

Figure 1. Eight categories of learners' differences

- Prior Knowledge
  - o Prior Knowledge and Achievement
  - o Structural Knowledge
- Demography
  - o Age and Generation*
  - o Gender
- Personality
  - o How do you allocate attention and act on or with information?
    1. Anxiety
    2. Tolerance for Ambiguity
    3. Tolerance for Frustration
    4. Tolerance for Unrealistic Experiences
  - o What do you want from experiences and information?
    1. Cautiousness
    2. Introversion vs. Extroversion
    3. Locus of Control
    4. Motivation
- Learning Styles
  - o Dunn and Dunn's Learning Styles
  - o Grasha-Reichman Learning Styles
  - o Gregorc Learning Styles
  - o Hill's Cognitive Style Mapping
  - o Honey and Mumford's Learning Style Questionnaire
  - o Kolb's Learning Styles*
  - o Salton's I-Opt
- Cognitive Styles
  - o How do you collect information?
    1. Haptic vs. Visual*
    2. Leveling vs. Sharpening
    3. Verbalizer vs. Visualizer
  - o How do you construct information?
    1. Conceptual Style
    2. Holist vs. Serialist
- Cognitive Controls: What are your abilities to collect, construct, and use information?
  - o Category Width
  - o Cognitive Complexity vs. Cognitive Simplicity
  - o Cognitive Flexibility
  - o Field Dependence vs. Field Independence*
  - o Focusing vs. Scanning
  - o Impulsivity vs. Reflectivity
  - o Strong Automation vs. Weak Automation
- Constituent Mental Abilities
  - o Contents
  - o Operations
  - o Products
- Common Mental Abilities/Intelligences
  - o Crystallized*, Fluid*, and/or Spatial

*Covered next in this chapter

Adapted from Jonassen, D.H., & Grabowski, B.L. (1993). *Individual differences, learning, & instruction.* Hillsdale, NJ: Lawrence Erlbaum.

than Millennials, but Schaie (1994) suggests that Baby Boomers, Gen-Xers, and Millennials differ in more than their attitudes and awareness.

Schaie (1994) surveyed his subjects in six abilities—inductive reasoning, numeric ability, perceptual speed, spatial orientation, verbal ability, and verbal memory. Subjects seemed to increase in inductive reasoning, spatial orientation, verbal ability, and verbal memory and decrease in numeric skills and perceptual speeds between about 25 and 65 years old. By 90 years of age, subjects showed stability in verbal ability and slumped in inductive reasoning, numeric ability, perceptual speed, spatial orientation, and verbal memory. Schaie identified several signs thought to slow such late-in-life losses, including above-average education and employment, adaptation to changes in midlife, satisfaction in midlife, satisfying and stimulating extracurricular activities, smart spouses, and stable perceptual speeds.

Similarly, Schaie (1994) split his subjects by their birthdays to show summative strengths and weaknesses in those same six abilities. Subjects born between 1907 and 1931 seemed stronger in numeric ability and weaker in inductive reasoning, perceptual speed, spatial orientation, and verbal memory than those born after 1931. Strangely, subjects born between 1924 and 1960 seemed stronger in perceptual speed and verbal ability than those born before 1924 and after 1960. Schaie suggested that such shifts in strengths and weaknesses were caused by increases in learners sent to primary, secondary, and postsecondary schools since 1931.

Do Millennials, Gen-Xers, and Baby Boomers show strengths and weaknesses similar to those suggested by Schaie (1994)? If so, they should show strengths in inductive reasoning, spatial orientation, and verbal memory and weaknesses in numeric ability. Millennials and Gen-Xers should show weaknesses in perceptual speed and verbal ability. Remember, Millennials are comfortable with PCs and PlayStations, two technologies that require perceptual speed and spatial orientation (Oblinger, 2003). Most likely, Millennials have higher hand–eye coordination, perceptual speed, and spatial orientation than Baby Boomers and Gen-Xers, suggesting that Millennials should seem more satisfied and skilled with technology-tinged distance education than Baby Boomers and Gen-Xers.

Differences in Learning Styles: Kolb's Learning Style Inventory

Lynn Curry, coauthor of *Integrating Concepts of Cognitive or Learning Style: A Review With Attention to Psychometric Standards—The Curry Report*, lamented a long-standing limitation of learning styles in an interview with *Training* in 2002: "One of the most pervasive difficulties in this field of research is the sloppiness of the definitions. Across the literature the same or similar concepts carry various titles, describe varying ranges of behavior and are very observable" (Delahoussaye, 2002, p. 32). Indeed, learning style tests and theories tally to about a dozen, including Dunn and Dunn's Model, Grasha-Reichman's Student Learning Styles Scale, Gregorc's Learning Style Delineator, Hill's Cognitive Style Mapping, Honey and Mumford's Learning Style Questionnaire, Kolb's Learning Style Inventory, and Salton's I-Opt. Here, we will review only Kolb's long-standing and long-used Learning Style Inventory.

Kolb's (1976) Learning Style Inventory ranks strengths and weaknesses in four abilities—concrete experience (CE), reflective observation (RO), abstract conceptualization (AC), and active experimentation (AE)—that were taken from his explanation of experiential education, which suggests that learners shift their strengths from one ability to another in their educational experiences. Of course, learners are strong in some abilities and weak in other abilities, particularly in abilities that are bipolar such as CE/AC and RO/AE. The Learning Style Inventory surveys learners' strengths and weaknesses in both bipolar constructs with four-phrase scales on which learners select which words are most similar and least similar to how they think they learn. Learners tally their scores from nine similarly set-up scales to learn their learning styles—converger, diverger, assimilator, and accommodator (Table 1). The activity for Thought and Reflection is provided to help you think about the predictive validity of the Kolb's Learning Style Inventory.

Table 1. Kolb's learning styles

Learning Style	Ability Areas
Converger	AC, AE
Diverger	CE, RO
Assimilator	AC, RO
Accommodator	CE, AE
Source: Kolb (1976)	

Activity for Thought and Reflection

Does Kolb's Learning Style Inventory have predictive validity?

Can Kolb's Learning Style Inventory predict your major or profession? Kolb's (1976) answer to that question recalled the riddle, "Which came first -- the chicken or the egg?" He pointed out that people pick majors or professions that match their learning styles and that their learning styles are strengthened by their picked majors or professions. Still, he suggested that students' learning styles point to particular professions:

- Converger•   Engineer
- Diverger•   Counselor/Manager
- Assimilator •   Scientist/Mathematician
- Accommodator•   Businessman or -woman

We checked out his theory unscientifically by testing a learning style with Kolb's Learning Style Inventory. This person scored high in Abstract Conceptualization and Reflective Observation, two abilities tied to Assimilators. If he is an Assimilator, Kolb's prediction suggests that he should be seeking a master's in biology, chemistry, mathematics, or physics, not agricultural education. Is he an anomaly?

In 1976, Kolb stressed that researchers and instructors should not select or split up learners using his test or theory. Rather, he suggested that they use his test and theory to talk about learners' styles with those learners. In 2002, Kolb repeated that remark:

> Few, if any, individual-difference tests can measure an individual with complete accuracy. For this reason, the Learning Style Inventory (LSI) is not recommended as a tool for individual selection purposes.... For this reason, we do not refer to the LSI as a test but rather an experience in understanding how you learn. (Delahoussaye, 2002, p. 30)

Do learners shift their learning styles after taking technology-taught courses at a distance? Can learners, whether at a distance or in a classroom, shift their learning styles at all? Kolb (1976) stresses that reliability, or stability, of learning styles seems shallow because learning styles are somewhat situational, suggesting that learners select or somehow shift their styles of learning from situation to situation. If so, learners should adapt to and learn from different situations and instructors. Researchers, learners, and instructors can use learning style tests and theories to reflect on and talk about learners' differences (Kolb, 1976; Jonassen & Grabowski, 1993; Delahoussaye, 2002)—not to shape research or styles of instruction.

Cognitive Styles: Perception

Abilities, characteristics, habits, patterns, perceptions, properties, propensities. Messick (1984) complained of the confusion of the countless conceptualizations of cognitive styles, calling researchers on their differing definitions and incorrect instruments. "As a consequence, style research is peppered with unstable and inconsistent findings (sic), while style theory seems either vague in glossing over inconsistencies or confused in stressing differentiated features selectively" (p. 59). He reviewed several slightly different conceptualizations of cognitive styles, to which he added his own, "characteristic self-consistencies in information processing that develop in congenial ways around underlying personality trends" (p. 61). Messick's conceptualization connected personality to styles of perception—aural, haptic, interactive, kinesthetic, olfactory, print, and visual (Table 2).

James and Galbraith (1985) recommend that educators match instruction to their learners' styles of perception. Surely, learners in a classroom or at a distance differ in how they perceive what they are presented, so James and Galbraith suggest that instructors make or take lessons and materials that match several of their learners' styles of perception. Rita Dunn, cocreator of Dunn and Dunn's model, recommends something similar to that of James and Galbraith. Instructors should survey their learners' styles, teach or train to those learners' highest-scoring styles, reteach or review to those learners' secondary styles, and sum up with their learners' creative constructions—paintings, poems, or songs—of what they have been taught (Filipczak, 1995). James, Galbraith, and Dunn recommend that instruction be directed to the most popular and prevalent styles, suggesting that educators reject or slight learners with less popular and prevalent styles of perception—and less popular and prevalent personalities (Messick, 1984). Of course, James, Galbraith, and Dunn did not intend that implication to be drawn from their recommendations, but their recommenda-

Table 2. How do you collect, perceive, or take in information?

Style	Sense	Strategies
Aural	Hearing	Aural learners best collect information from lectures and by listening to others.
Haptic	Touch	Haptic learners see with their hands, not with their eyes. They best collect information by feeling, holding, or touching things linked to or include that information.
Interactive	Speech	Interactive learners best collect information by debating with and talking to others about that information.
Kinesthetic	Movement	Kinesthetic learners are movers and shakers. They best collect information by bouncing, rocking, twitching, or walking while they collect that information.
Olfactory	Smell and Taste	Olfactory learners best collect and construct information by tying it to smells or tastes taken in while collecting that information.
Print	Sight	Print learners best collect and construct information by reading and writing about that information.
Visual	Sight	Visual learners are "show me" learners. They best collect information by looking at charts, diagrams, and pictures, and observing others work.

Source: James & Galbraith (1985)

tions did recall the one-size-fits-all strategy on instruction to which educators and trainers tend to surrender themselves.

How should learners with such styles of perception react to rejection of their styles and still succeed in distance education? In 1995, Bill Filipczak, then staff editor of *Training*, offered his readers instructions on how to learn from his print presentation of styles of perception. He recommended that auditory, or aural, learners read his article aloud; kinesthetic learners read his article and pace; and tactual, or haptic, learners read his article with highlighters in hands. Of course, his recommendations seem simple and tongue in cheek, but Filipczak's recommendations seemed smart and true too, suggesting that learners might make lessons and materials match their styles of perception to make up for limitations of or mismatches made in distance education.

Cognitive Controls:
Field Dependence/Independence

Which cognitive construct is subordinate to the other—cognitive styles or cognitive controls? Messick (1984) compared and contrasted the constructs, calling the controls subordinate to the styles. Jonassen and Grabowski (1993) reversed the order made by Messick, setting the styles subordinate to the controls, or permanent patterns of thought that tailor perception and personality to the person. Here, we will refer to their conceptualization of cognitive controls and relate distance education to field dependence/independence (Table 3), which suggests that learners split on their answers to the question, "Is the whole greater than the sum of its parts?"

How should instructors design distance education environments that correspond to field dependents and independents? Joughin (1992) lists four elements of education—acceptance and source of authority, analytical ability, source of social support, and source of structure—that differ from field dependents and independents and shape their self-direction, an element important to distance education: "Consequently, adult educators need to be constantly aware of these elements and the differing implications they will have for learners with differing cognitive styles" (p. 13). Surely, instructors should

Table 3. Learning preferences and behaviors of field dependents vs. field independents

Field Dependents	Field Independents
• Like group-oriented and collaborative learning • Prefer clear structure and organization of material • Attend to the social components of the environment • Respond well to external reinforcers • Prefer external guidance	• Like problem solving • Prefer situations in which they have to figure out the underlying organization of information (e.g., outlining) • Like transferring knowledge to novel situations • Prefer independent, contract-oriented learning environments • Respond well to inquiry and discovery learning
Source: Knowles, Holton, & Swanson (1998, p. 160)	

survey their learners' field dependence/independence preferences and refer to the surveys to structure their lessons and materials? But learners, too, should see, seek, and support their analytical abilities, sources of authority, sources of social support, and sources of structure to improve their self-direction in, satisfaction with, and success in distance education.

Intelligences:
Crystallized and Fluid Intelligences

Constructions of cognitive controls, cognitive styles, and learning styles suggest that a learner with this control or style is more intelligent than another learner with a different control or style. Sternberg (1997) rejects a similar statement in his study of intelligence, in which he defines the capstone cognitive construct. To Sternberg, intelligent acts and behaviors seem situational, or summations of biology, culture, environment, and time. Similarly, Cattell (1963) conceptualizes two types of intelligence that are caused by and change with biology, culture, environment, and time—crystallized intelligence and fluid intelligence. Here, we will review the conceptualization created by Cattell and return to the definition of intelligence described by Sternberg.

How do you define "crystallized"? Solid? Stable? Unchanging? Cattell's (1963) connotation of "crystallized" suggests the summation, not subtraction, of educational and occupational skills that were sought and strengthened with time. "Crystallized ability loads more highly those cognitive performances in which skilled judgment habits have become crystallized (hence its name) as a result of earlier learning application of some prior, more fundamental general ability to these fields" (p. 3). He cautions that crystallized intelligence changes with content and culture, not age, generation, or time, and concludes that crystallized intelligence helps learners strengthen or support their skills. Similarly, Cattell cautions that fluid intelligence changes with culture and fails with acuity and age, similar to what was suggested by Schaie (1994), and concludes that fluid intelligence helps learners study new skills and try new tricks, a conclusion that complements his conceptualization of fluid intelligence. "Fluid general ability, on the other hand, shows more in tests requiring adaptation to new situations, where crystallized skills are of no particular advantage" (p. 3).

Which type of intelligence—crystallized or fluid—helps or hinders success in distance education? Neither. Cattell (1963) surveyed middle schoolers' intelligence and successes in school and surmised that their successes were supported by their crystallized intelligence and fluid intelligence—and their means of motivation and personalities. In this regard, adults probably do not differ significantly from middle schoolers. If so, intelligence seems a dispositional and situational summation of biology, culture, environment, and time (Sternberg, 1997) that can be called on, changed, and combined to seek success in distance education.

Which Difference Matters Most to You and Your Learning?

Clearly, those who learn differ in how they learn, whether those differences are derived from demography, learning styles or cognitive styles, controls, and intelligences. But one difference, part personality, part responsibility, and part situation, matters more to you and your learning than its counterparts—motivation for achievement or edification. Ashley Fields of Shell Oil Company's Shell People Services expressed in an interview with *Training* in 2002: "Learning watchmaking is a lot different than learning to be a commodity trader or learning hands-on tasks vs. knowledge-based tasks. Yet, if provided sufficient incentive, say $1 million, most people can learn to do both" (Delahoussaye, 2002, p. 34). Here, Fields recommends an extrinsic reward to motivate learners, not an optimal setting for learner success.

Given that most instructors do not have $1 million to motivate learners, additional strategies are needed. Mandigo and Holt's (2002) OPTIMAL strategies model (discussed in greater length in chapter V), provides a good framework from which such strategies can be considered.

- *Opportunity for success:* Instructors should offer their learners opportunities to succeed by suggesting that learners set self-competitive—not class-competitive—goals.

- *Perception of choice:* Instructors should offer their learners the choice to change an activity or create an alternative activity.

- *Task mastery:* Instructor should suggest and support learners' strategies to learn tasks for the tasks, not for their competitions, egos, or popularity.

- *Inclusion teaching style:* Instructors should encourage learners to enter activities or assignments at their levels of ability.

- *Motivate through intrinsic elements:* Instructors should replace external evaluations and rewards with learners' self-evaluations of their activities and assignments.

- *Abilities awareness:* Instructors should teach to several levels of skill.

- *Like to do it:* Instructors should ask about and assign activities and assignments liked by their learners.

Distance learners should support themselves and their instructors and take responsibility for their differences, educations, and means of motivation ... by reading their textbooks. Are you still reading? Ah ... motivation!

Conclusion

This chapter focused on principles of adult learning and learner differences. We believe that adult learning theory/principles should be considered for effective instruction and learning at a distance. We highlighted age and generation, Kolb's Learning Styles Inventory, cognitive styles (haptic vs. visual), field dependence versus independence, and mental abilities/intelligences (crystallized vs. fluid) to illustrate this point. While the instructor has a responsibility for creating the learning objects and facilitating instruction, ultimately it is the learners' responsibility to engage, manage course content/activities, and complete assignments. Therefore, the primary role of an educator or trainer when instructing at a distance is that of the coach, mentor, or facilitator. If educators and trainers consider the myriad of learner differences and focus of the use of the OPTIMAL strategies and principles of andragogy, more meaningful learning can occur. In the next chapter, we shall examine how to engage adult learners and foster self-directedness.

Application Exercise

Take one of the learning styles inventories below (Kiersey
Temperament Sorter, OVARKA Learning Styles Inventory, OR one
of your own choosing). What were your results? Does the
description of your profile describe you? How does it differ? What
"personality" or "learning preference" does it neglect that may be
necessary to describe learning characteristics? Does it match with
your philosophy of education?

http://www.advisorteam.com/temperament_sorter/register.asp?partid
http://www.aged.tamu.edu/research/projects/tests/Ovarka/

References

Bennett, C., & Rockwell, K. (1995). *Targeting outcomes of programs
(TOP): An integrated approach to planning and evaluation.* Unpublished manuscript. Lincoln: University of Nebraska.

Brookfield, S.D. (1986). *Understanding and facilitating adult learning: A
comprehensive analysis of principles and effective practices.* San
Francisco: Jossey-Bass.

Cattell, R.B. (1963). Theory of fluid and crystallized intelligence: A critical
experiment. *Journal of Educational Psychology, 54,* 1-22.

Delahoussaye, M. (2002). The perfect learner: An expert debate on learning
styles. *Training, 39,* 28-36.

Dewey, J. (1938). *Experience and education.* New York: Collier Books.

Filipczak, B. (1995). Different strokes: Learning styles in the classroom.
Training, 32, 43-48.

Fulford, C., & Zhang, S. (1993). Perceptions of interaction: The critical
predictor in distance education. *The American Journal of Distance
Education, 7*(3), 8-21.

Garrison, R. (1990). An analysis and evaluation of audio teleconferencing to
facilitate education at a distance. *The American Journal of Distance
Education, 4*(3), 13-24.

Howard, C., Schenk, K., & Discenza, R. (2004). *Distance learning and university effectiveness: Changing educational paradigms for online learning.* Hershey, PA: Idea Group.

James, W.B., & Galbraith, M.W. (1985). Perceptual learning styles: Implications and techniques for the practitioner. *Lifelong Learning, 8,* 20-23.

Jonassen, D.H., & Grabowski, B.L. (1993). *Individual differences, learning, & instruction.* Hillsdale, NJ: Lawrence Erlbaum.

Joughin, G. (1992). Cognitive style and adult learning principles. *International Journal of Lifelong Education, 11,* 3-14.

Knowles, M.S. (1990). *The adult learner: A neglected species.* Houston, TX: Gulf.

Knowles, M.S. (1995). *Designs for adult learning: Practical resources, exercises, and course outlines from the father of adult learning.* Alexandria, VA: ASTD.

Knowles, M.S., Holton, E.F., III., & Swanson, R.A. (1998). *The adult learner: The definitive classic in adult education and human resource development.* Woburn, MA: Butterworth-Heinemann.

Kolb, D.A. (1976). *The learning style inventory.* Boston: McBer.

Leamnson, R. (1999). *Thinking about teaching and learning: Developing habits of learning with first year college and university students.* Sterling, VA: Stylus.

Lindner, J.R., Dooley, K.E., & Williams, J.R. (2003). Teaching, coaching, mentoring, facilitating, motivating, directing … What is a teacher to do? *The Agricultural Education Magazine, 76*(2), 26-27.

Mandigo, J.L., & Holt, N.L. (2002). Putting theory into practice: Enhancing motivation through OPTIMAL strategies. *Avante, 8*(3), 21-29.

McKeachie, W.J. (1999). *Teaching tips: Strategies, research, and theory for college and university teachers* (10th ed.). Boston: Houghton Mifflin.

Messick, S. (1984). The nature of cognitive styles: Problems and promise in educational practice. *Educational Psychologist, 19,* 59-74.

Moore, M. (1980). Independent study. In R. Boyd, & J. Apps (Eds.), *Redefining the discipline of adult education* (pp. 16-31). San Francisco: Jossey-Bass.

Moore, M., & Kearsley, G. (1996). *Distance education: A systems view*. Belmont, CA: Wadsworth.

Oblinger, D. (2003). Boomers, gen-xers, and millennials: Understanding the "new students." *EDUCAUSE Review, 38,* 36-40.

Pratt, E.D. (1988). Andragogy as a relational construct. *Adult Education Quarterly, 38,* 160-161.

Richards, L.J., Dooley, K.E., & Lindner, J.R. (2004). Online course design principles. In C. Howard, K. Schenk, & R. Discenza (Eds.), *Distance learning and university effectiveness: Changing education paradigms for online learning* (pp. 99-118). Hershey, PA: Idea Group.

Ritchie, H., & Newby, T. (1989). Three types of interaction. *The American Journal of Distance Education, 3*(3), 36-45.

Sabini, J., Siepmann, M., & Stein, J. (2001). The really fundamental attribution error in social psychological research. *Psychological Inquiry, 12*(1), 1-15.

Schaie, K. (1994). The course of adult intellectual development. *American Psychologist, 49,* 304-313.

Sternberg, R.J. (1997). The concept of intelligence and its role in lifelong learning and success. *American Psychologist, 52,* 1030-1037.

Chapter V

Engaging Adult Learners and Fostering Self-Directedness

 Making Connections

It is clear that educators rely on a variety of instructional methods to change learners' behaviors. What is less clear is how distance learning educators can foster deeper and more meaningful learning by taking into account a learner's unique background, experiences, competencies, learning styles, personality type, and levels of self-directedness. This is a challenge for those educators wishing to create a learner-centered instructional environment at a distance. How can educators avoid the trap of "teaching to the middle," providing materials that are too challenging for some learners and too simple for others? How can educators identify learners' dependency level/self-directedness?

Introduction

There are many benefits of distance education enjoyed by learners and educators. Some benefits are more obvious than others. An obvious benefit, for example, is that learners that are time and/or place bound can benefit by taking courses at times and places that are convenient for the learner. A less obvious benefit may be the ability of the instructor to tailor the instructional sequence to take advantage of the learner's unique competencies (Dooley & Lindner, 2002; Lindner & Dooley, 2002). Consider the following example. For a particular course the instructor has developed two learning tracks. The first track allows learners, who prefer more direct experiences, to complete authentic learning activities related to the content. The second track allows learners, who prefer more abstract experiences, to read and write about the content. By providing learners the option to navigate the course, based on their unique competencies, the instructor maximizes the potential for learning. To often, instruction offered at a distance does not take into account learners' unique competencies, resulting in learners that are not actively engaged and learners that are unnecessarily dependent on the instructor (Lindner, Dooley, & Murphy, 2001). Some of the issues addressed in this chapter that affect learning, engagement, and self-directedness include learner temperament and personality, gender, attrition, learner responsibilities, rigor, satisfaction, quality, delivery strategies, and the role of the educator. We will provide a brief overview of some of the factors that impact learning in distance education settings, recognizing that much more research is needed to fully understand how to maximize learning at a distance.

Learner Characteristics Impacting Distance Learning

The possibility that distance education learners' personality may play a role in predicting satisfaction and achievement has resulted in numerous studies. While some research suggests differences in learner satisfaction with distance education courses based on the learners' personality type (Daughenbaugh, Ensminger, Frederick, & Surry, 2002) other research suggests that learners' temperament does not have a major impact on satisfaction or learning outcomes (Stokes,

2001). The influence of distance education students' personality and learning, engagement, and self-directedness has not been determined.

Gender has been identified as a factor that may affect learning, engagement, and self-directedness of distance education learners. In a study that compared female to male students' progress through an asynchronously delivered course, Lindner, Dooley, and Hynes (2003) found that engagement was related to gender. In this study, it was reported that females tended to engage in the course sooner and complete the course sooner than did males in the course. A basis for this finding was not determined. These authors noted that although females and males engaged differently in the course, learning outcomes did not differ. Krikup and von Prummer (1990) noted that in distance education courses, female students interacted more with the instructor and other students than did male students. Krikup and Prummer (1990) and Lindner, Dooley, and Hynes (2003) recommended that instructors take into account these differences when developing and delivering distance education courses.

Attrition of distance education learners is a major concern in distance education (Wickersham & Dooley, 2001). While the exact reasons for learner attrition are not known, Saba (2000) recommended that instructors need to focus on helping learners be more self-directed, helping learners manage their time better, helping learners create learning environments, maintaining good communications and responding in a timely manner to learners' emails, and using learning communities. Knowles, Holton, and Swanson (1998) noted that adult learners' need to have a reason to learn. For example, an adult may have limited opportunities career advancement because of insufficient skills. If the learner desires career advancement, motivation to learn new skills will be higher. Knowles (1990) theory on adult learning suggests that learners are self-directed. Given Knowles assumptions, Merriam (2001) noted that attrition in distance education should be expected, because self-directed learning will have learners beginning, pausing, stopping, and completing instruction units as needed. Pappas, Lederman, and Broadbent (2001) recommended that distance education instructors' need to monitor learners' initial engagement, progress throughout the course, and performance to improve learners' completion rates and to identify those learners at risk for note completing. Course design features that can promote engagement and interaction that minimize attrition are discussed later in this chapter.

Another way to address the problem of attrition is to better understand the challenges facing distance education learners and strategies that can be used to

Thought and Reflection

The field of adult education has long embraced such ideas as autonomy, independence, and personal development of adult learners. These ideas are implicit in such terms as lifelong learning, self-directed learning, self-planned learning, independent study, distance education, learning projects, andragogy, and self-directed learning readiness. All of these in some way stress the role of individual learners in the learning process. What is the role of the learner?
Source: Brockett & Hiemstra, 1991

help the learner overcome problems that arise. Howland and Moore (2002) noted learners must ultimately take responsibility for their own learning. Failure of distance learners to take such responsibility as managing time effectively may result in dissatisfaction and attrition (Weinstein, 2002). Poor communication skills between instructors and learners become more problematic in distance education courses than in face-to-face courses (Lindner & Murphy, 2001; Miller & Pilcher, 2000).

Interactions and Engagement

The instructors' role in maximizing learner interactions and engagement are necessary for distance education learners to be successful (Kearsley & Shneiderman, 1999). Research has shown that a learner's length of engagement in an asynchronously delivered course is positively related to that student's success in the course (Lindner, Hynes, Murphy, Dooley, & Buford). Taplin, Yum, Jegede, Fan, and Chan (2001) noted that maximizing learner interactions and engagement in a distance education course is different than that in a face-to-face course. While instructors can use strategies such as cohorts to stimulate interactions and engagement, doing so at a distance requires more planning and effort by instructors than in face-to-face courses (Brown, 2001).

In chapter I, we provided a basic definition of distance education. For the purpose of discussing the importance of maximing interaction in a distance education environment, we offer the following expanded version. Distance education is the application of delivery strategies using a variety of delivery methods with learners who are constrained by time/space/lifestyle. Throughout this book we emphasize the notion that delivery methods do not matter regardless of whether a course is taught face-to-face or over interactive/ compressed video (ITV), synchronous or asynchronous, Internet protocol (IP) conferencing, or streaming video. Learners can be successful and few, if any, differences in performance will be found by type of delivery method. While delivery methods do not matter, delivery strategies do. Delivery strategies can be defined as those methods used to engage learners in the instructional materials. It can be argued that through various interactions, engagement results in learning … not grades (Merrow, 2003).

Moore (1989) developed a model for describing interactions used to engage learners in a distance education environment. His model includes three types of interaction: (1) learner-to-learner; (2) learner-to-content; and (3) learner-to-instructor. Hillman, Willis, and Gunawardena (1994) expanded this model to include an additional type of interaction: learner-to-technology (interface). Moore (1989) stated that "educators need to organize programs to ensure maximum effectiveness of each type of interaction..." (p. 5). Fulford and Zhang (1993) stated that "the critical predictor of satisfaction is the perception of overall or vicarious interaction" (p. 17).

Examples of learner-to-learner interactions include online chats, threaded discussion, e-mail, point-to-point video conference, audio calls, and so forth. Examples of learner-to-content interactions include online books, online instructional materials, support materials, worksheets, case studies, and so forth. Examples of learner-to-instructor interactions include lecture, e-mail, online editing and feedback, evaluation of learning, interactive television, streaming video, voice-over PowerPoints©, and so forth. Examples of learner-to-technology interactions include online tutorials on how to use educational teachnology, getting help online, downloading plug-ins, installing software, file management including uploading and downloading files, electronic libraries, and so forth.

The model presented (Figure 1) shows the interplay among the interactions (Figure 1) when overlapping of interactions occurs. To maximize learning and increase satisfaction in a distance education environment, learner-to-learner, learner-to-content, learner-to-instructor, and learner-to-technology interac-

Figure 1. Depiction of vicarious interaction and maximized learning and satisfaction resulting from four learner relationships

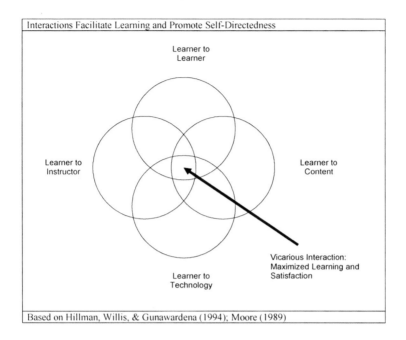

tions should be included. Along with facilitating deeper and more meaningful learning, maximizing interactions also promotes self-directedness among learners.

A Question of Quality

Key questions of interest with respect to face-to-face and distance education that continue to arise among educators, learners, and trainers include the following: Are face-to-face educational opportunities more academically rigorous than distance educational opportunities? Are learners at a distance as satisfied with their educational experience as face-to-face learners? Is the quality of the learning experience the same at a distance as it is face-to-face? Are synchronous delivery strategies more effective than asynchronous delivery strategies?

As noted in chapter II, Russell (1999) documented that learner outcomes are the same regardless of whether the educational opportunity is face-to-face or at a distance. Clark (1994) argues that when the rigor in development and delivery of face-to-face and distance educational opportunities are the same, learner outcomes will also be the same. The area of concern, therefore, is that distance education lacks rigor. Miller and Shih (1999) noted that active learning, effort, and high cognitive levels are factors useful in explaining educational rigor. To ensure rigor, distance educators should account for these factors when developing, delivering, and evaluating distance education offerings. These authors suggest that if distance education offerings are less rigorous than face-to-face offerings, distance educators should seek means to engage the learner further, increase the amount of effort the learner must put forth, and authenticate learning at higher levels of cognition.

In addition to ensuring rigor in distance educational opportunities, the quality of the educational experience must also be addressed. Giguere and Minotti (2003) suggest guidelines for ensuring high-quality educational opportunities. These guidelines include "Learner-centered curriculum; pedagogically [we suggest *andragogically*] appropriate activities; clearly defined objectives, goals, and expectations; easily accessible content; content in multiple formats; sense of community; shorter, focused training; expert online facilitation; and immediate online and off-line technical support" (pp. 57–58). The focus should, therefore, be on measures of quality, not delivery methods. The extent to which a distance education (or face-to-face) opportunity is of high quality is not a function of comparing one to the other. We discuss throughout this book best practices which serve as high-quality measures.

Several studies (Dooley, Kelsey, & Lindner, 2003; Murphy, 2000) have shown that regardless of whether a learner is at a distance or face-to-face, overall satisfaction with the educational experience is similar. While technology is making it easier to conduct evaluations of the effectiveness of programs, courses, and educators, it is also introducing new biases not previously addressed. Because an instructor's effectiveness is often measured, reported, and evaluated based solely on a learner's responses on evaluation forms, it is necessary to balance what we can do with what will provide us with normalized and useful data. A major concern is that of using electronic forms in distance courses versus using traditional paper forms to evaluate educators' effectiveness in teaching, and then comparing scores without accounting for differences in how the course was delivered. Differences in the nature of questions asked must also be addressed.

We believe that the *ideal* learning environment includes face-to-face and real-time interactions between and among educators and learners. Such synchronous interactions, however, are not necessary for successful learner outcomes, rigor, and/or quality with respect to distance learning. Limitations to learning synchronously, face-to-face, include lack of convenience, loss of interpersonal interaction, and failure to develop real-world skills (Rea, White, McHaney, & Sanchez, 2000). Rea, White, McHaney, and Sanchez further suggest that these limitations to learning face-to-face, coupled with advances in information technologies, particularly the Internet, have contributed to the rapid adoption and diffusion of distance education.

While distance education has been successful in addressing certain limitations of face-to-face learning, learners at a distance face other inhibiting factors. In addition to feeling isolated and lonely, Kinshuk (2003) identifies five limitations to learning asynchronously at a distance: (1) lack of match between course material and its explanation; (2) lack of contextual discussion; (3) lack of human teacher expression and explanation; (4) lack of human interaction; and (5) lack of contextual understandings.

Although the literature is replete about what is the same between traditional learners and distance learners, there are, in fact, major differences in how these groups engage in the learning process.

Self-Directed Learning

In *30 Things We Know for Sure About Adult Learning*, Zemke and Zemke (1981) noted that "The adult learner is a very efficiency-minded individual" (p. 47). We subscribe to the ideal that distance learners are ultimately responsible for their own learning. Self-directed learning includes both internal characteristics of the learner and external characteristics of an educational process (Brookfield, 1986; Brockett & Hiemstra, 1991). The role of an educator should ideally vary from teaching, coaching, mentoring, facilitating, motivating, and directing, based on the learners' needs or capabilities (Lindner, Dooley, & Williams, 2003).

Fostering Deeper and More Meaningful Learning

To create learning environments that foster deeper and more meaningful learning, educators need to focus on the learners' motivation to learn, learner-centered instruction, and providing a setting that promotes engagement and interaction.

While taking into account a learner's unique competencies and characteristics may not result directly in deeper and more meaningful learning, it will provide a more comfortable learning environment in which learners can succeed.

What is the role of the distance educator in fostering deeper and more meaningful learning? Grow (1991) suggested that the teaching methods used by an educator should be based on the learners' level of self directedness. Grow's (1991) Staged-Self-Directed Learning (SSDL) model provides a good framework from which to consider the learner's and the educator's roles in the learning process. Key assumptions of the model are: Self-direction can be taught; self-direction can be learned; self-direction is a situational ability; an idealistic teaching style/method does not exist; and self-direction and creation of lifelong learners is an overarching educational goal.

There are four levels of educators' and learners' self-directedness in Grow's SSDL model (Figure 2). Grow used the nomenclature S1, S2, S3, and S4 to describe the learner's stage of dependence. T1, T2, T3, and T4 were used to describe stages of the educator. S1 learners are those who are completely dependent on the instructor. S2 learners are those who need external motivation by the instructor. S3 learners are those who are motivated to learn, but need to be empowered by the instructor. S4 learners are those that are self-directed and need minimal guidance from the instructor.

Learning is hampered when mismatches between learners and instructors occur (Figure 3). Grow (1991) noted that some mismatches have more severe

 **Internet Connection**
http://www.longleaf.net/ggrow/

This site provides an expanded description and discussion of Gerald Grow's Staged Self-Directed Learning Model.

Figure 2. Grow's (1991) Staged Self-Directed Learning Model

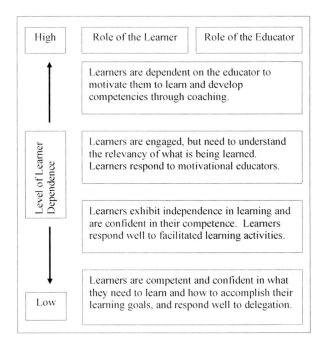

 Thought and Reflection

What Is Andragogy? An Anonymous Student's Critique

In 2002, John R. Rachal asked his readers, "What is andragogy?" He had read and regurgitated more than a dozen connotations, most of which were printed in distant dissertations, and one denotation, that of its starter and supporter, Malcolm Knowles, printed in Knowles' *The Modern Practice of Adult Education: From Andragogy to Pedagogy*. Of course, Knowles called andragogy the art and science of helping adult learners learn. Rachal referred to and wrote that denotation in his observation that his peers ponder and study its art—its philosophy and theory—and shaft its science—its empirical effectiveness in education. He linked that lack of study of its empirical effectiveness to its lack of an operational definition that is both consistent and correct. Rachal linked that lack of an operational definition of andragogy to something called "paradigm devolution," or an abasement of an ideal to an ideology by its so-called supporters. What do you think? What is andragogy?

Figure 3. Grow's Staged Self-Directed Learning mismatches

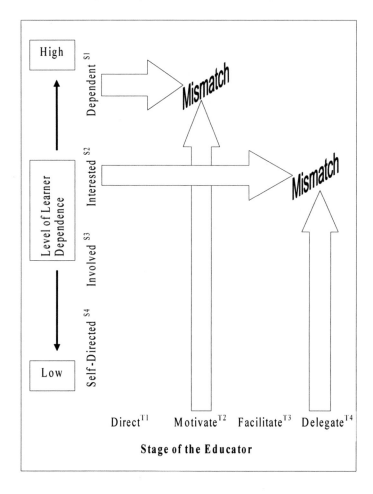

consequences than others. For example, the instructor who insists on maintaining full control over a learning environment, primarily comprised of self-directed students, will undoubtedly, have major problems motivating and engaging those students in the learning process.

Thought and Reflection

Consequences of Mismatches: An Anonymous Student's Reaction

Grow (1991) notes two mismatches that can cause severe problems for learners and educators. First, self-directed learners might rebel from or resent authoritarian educators because both learners and educators are vying for control over the learning process. Second, dependent learners might resent delegating educators because those learners might lack skills needed to take responsibility for their own learning. Grow suggests several uses of SSDL—in classes and curricula, linear and looping—but he cautions that its ease and elasticity causes him to question SSDL.

Tennant (1992) questions Grow's SSDL model, suggesting that it seems too simple and shallow to use. He names numerous inconsistent or incomplete constructions made by Grow. For example, he writes that Grow did not note at what time teachers should change their teaching to move their students toward self-direction. Tennant answers his own question with his suggestion that mismatched levels of learning and teaching might move students more than matched levels. Grow (1994) responds to Tennant's critique by agreeing that teachers might mismatch levels to move their students toward self-direction. Grow shares several examples, research, and uses of SSDL, suggesting that others seem satisfied with his simple and shallow study of student self-direction. But I [the anonymous student] am not satisfied with SSDL—I think matching, mixing, and moving from level to level loses learners.

Recently I took a class that was taught by an educational constructivist who wanted his students to move toward self-direction and love of lifelong learning. Some students stepped up and slid toward self-direction with success, but others stayed stationary and resented our teacher because they did not have—or want to have—those skills needed for self-direction. Our teacher lost several of his students because he mismatched their levels of learning—dependent, interested, and involved—to his levels of teaching—facilitator and consultant/delegator. I cannot and will not take that risk in my classroom.

Toward Self-Directed Learner

We believe that distance learners need to have higher levels of self-directedness than do traditional learners in order to be successful. This suggests that andragogical learner principles, discussed in Chapter IV, are more appropriate than pedagogical learner principles for distance education instruction. Many distance education instructors make the false assumption that all distance learners are, by the mere fact of signing up for instruction at a distance, self-directed. Our collective experiences of teaching at a distance would suggest otherwise. For example, our online learners range from dependent learners to self-directed learners. This presents unique challenges for those instructors attempting to individual instruction and maximize the learning experience. Not all learning experiences, however, can or should be individualized. If an instructor, for example, is conducting online training for a particularly large group of learners, the practicality of such is limited. When a distance education instructor is confronted with the example above, what can they do to maximize overall learning? We suggest instructors should start with a balance of pedagogical and andragogical teaching methods. By incorporating both directed learning activities (writing a one-page reaction paper to a required reading, e.g.) and self-directed learning activities (writing a reflection paper on personal experiences related to the content, e.g.), the instructor is more likely to facilitate learning for the entire class.

Implications for Educators

Identifying learners' level of self-directedness (Lindner, Dooley, & Williams, 2003) is difficult and as Grow (1991) suggests, situational. The instructor should rely on a variety of means to determine learners' level of self-directedness. Instructors' past experiences of working with similar learners should be considered. The following questions can be used to help guide an instructor in determining learners' level of self-directedness: how did the learners' respond to structured writing assignments? Did the students need repeated help to complete the assignment? Were learners able to jump in to threaded discussions or did they need help getting started? Were students able to run with self-guided projects or did they need assistance in selecting a topic?

Instructors might also use strategies such as providing examples/samples or sufficient details in course syllabi or assignment descriptions to help learners who are not as self-directed feel more self-assured. The loss of nonverbal cues and physical presence in online learning environments can be intimating to some learners. Instructor immediacy behaviors, such as using the learners' name and illustrations from the learners' work to make an instructional point will make learners feel included in the virtual learning community. As learners' become more confident and competent, their innate self-directedness and motivation are allowed to flourish.

 Internet Connection
http://home.twcny.rr.com/hiemstra/sdlindex.html

This site provides an excellent discussion of theory, research, and practice on self-direction. It also offers an annotated bibliography of seminal work with respect to andragogy and self-direction.

Enhancing Self-Directedness Through OPTIMAL Strategies

Learners have a need to solve problems regardless of external motivation, feedback, or reward (Bransford, Brown, & Cocking, 2000). This need for learning is called "competence motivation" or "achievement motivation." Harter's (1978) Competence Motivation Theory suggests that the more intrinsically motivated learners are, the more they perceive themselves as being competent. This theory is the basis for Mandigo and Holt's OPTIMAL Strategies model (2002). We believe this model has practical applications in distance education for enhancing a learner's internal motivation or self-directedness. The model is comprised of seven instructional strategies.

Opportunities. Educators should provide distance learners with a variety of learning opportunities or engaging assignments early in the course. Research on patterns of engagement found that, in fact, early engagement in a distance education experience was related to a learner's overall success and satisfaction (Lindner, Dooley, & Hynes, 2003; Lindner, Hynes, Murphy, Dooley, & Buford, 2003).

Perceptions. Learners' enjoyment and competence is higher when they have opportunities to make choices with respect to what they will do, how they will do it, and how it will be evaluated. For self-directed learners, this is a particularly important strategy; for directed learners, it is less important (referring back to our discussion of Grow's SSDL model). A learning activity that supports perceptions would be a self-directed application project. For example, you could ask learners to develop a training module for other adult learners, and encourage them to make it as applicable and contextual as possible. You could suggest major items in the module that the learners could include, if they so desired, and indicate that the final format and length of the module will vary dramatically depending on what the learners want to achieve. Such a project will afford the self-directed learner with several opportunities to make choices and will provide some structure for the directed learner.

Task. Learning opportunities that are task oriented contribute more to positive learning behaviors than do ego-oriented learning opportunities (Ames, 1992). Task-oriented learning opportunities include those that focus on self-improvement and acquisition of new competencies. Ego-oriented learning opportunities focus on competition among learners and comparisons against rigid standards. Learning environments that give learners flexibility in choosing what to work on and when to work on it are task oriented. Distance education at its core should be task oriented. The following example of task orientation is drawn from one of our graduate classes on adult learning theory. In this course we inform students that the course is designed to be asynchronously delivered, meaning they can work on meeting the stated objectives of the course at any time or location (within specific time goals). They can also work on most assignments out of sequence. For example, they can work on Modules 1 and 4 before working on Modules 2 and 3.

An example of ego orientation that is carried out every day is assessing learner success against rigid standards or grading on the bell curve. In their controversial book, Herrnstein and Murray (1994) conclude that human intelligence is normally distributed along a continuum that is bell-shaped. Such assessment of learner achievement, we would argue, is narrow-minded and counterproductive to the learning process, and may, in fact, impede learning.

Inclusion. The increasing heterogeneity of distance education learners poses a concern for educators. For educators who are used to working with homogeneous groups, distance education learners' varied backgrounds, demographics, motivations, goals, personal preferences, personalities, and experiences often result in higher levels of frustration between and among educators and learners when inclusive teaching methods are not used. The Center for Instructional Development and Research (2003) suggests the following six strategies for inclusive teaching: (1) convey respect, fairness, and high expectations; (2) consider learner differences; (3) support learner success; (4) foster equitable participation; (5) diversify your instructional style; and (6) transform the curriculum.

Motivation. Learner motivation can be divided into two categories: extrinsic motivation and intrinsic motivation (Sansone & Harackiewicz, 2000). Extrinsic motivation is derived from learners' responses to forces beyond themselves. Extrinsically motivated people respond to reward, praise, good grades, money, and so forth. Intrinsic motivation is derived from learners' internal drive. Intrinsically motivated people want to learn for the sake of learning, just to know something new, and out of curiosity. Lumsden (1994) suggests that educators should first seek to motivate learners intrinsically. This author further noted that motivating learners extrinsically may actually undermine their ability to be intrinsically motivated. The very nature of distance learners suggests that learners tend to be more intrinsically motivated.

Abilities. Talent on task is a powerful concept. Self-assessments, such as those described in chapter I, are critical to documenting a distance learner's abilities to determine if he or she is more confident than competent. Having this information is necessary for an educator to develop individualized learning sequences needed to optimize learning. Such a process allows the instructor to match the task to the learner, and not the learner to the task.

Likeability. To keep distance learners on task and challenged, educational strategies that focus on enjoyment and interest are needed. Such strategies include using learner-centered instructional environments in which achievement for the distance learner is maximized. Inquiry-based learning strategies and contextually rich curricular materials should be used.

 Internet Connection
http://depts.washington.edu/cidrweb/inclusive

This site provides an excellent discussion of inclusive teaching.

Conclusions

In this chapter, we reviewed ways to engage the distance learner and how to foster the learner's self-directedness. Major themes of the chapter include the following: Learner's personal characteristics should be accounted for in the teaching and learning process. Distance learners are responsible, ultimately, for their own success or failure. Learning for understanding can be enhanced through vicarious interactions between and among learners, educators, technology, and content. Academic rigor, learner satisfaction, high-quality educational experiences, and delivery methods must be addressed by the educator. Educators have an important role in developing and fostering learners' self-directedness. Use of the seven teaching strategies discussed may enhance learners' self-directedness. When melded with the theory of andragogy, models of learning, learner self-directedness, engagement, and interactions provide a powerful framework from which to design and deliver educational programs and courses at a distance. In the next section of this book, we will bring all these concepts together using systematic instructional design.

 Application Exercise

Identify a teaching situation in which you provided too much or too little direction for a group of learners. Within the framework of Grow's SSDL model, describe how the learners, in general, responded to this level of direction. Why do you think the learners responded in the manner in which they did? Could you have avoided ALL mismatches between the learner and yourself? What could you have done to help the learners be more self-directed? Is it the educator's role to help learners be more self-directed?

References

Ames, C. (1992). Classrooms: Goals, structures, and student motivation. *Journal of Educational Psychology, 84*, 261-271.

Bransford, J.D., Brown, A.L., & Cocking, R.R. (2000). *How people learn. Brain, mind, experience, and school*. Washington, DC: National Academy Press.

Brockett, R.G., & Hiemstra, R. (1991). *Self-direction in adult learning: Perspectives on theory, research, and practice*. New York: Routledge. Retrieved January 7, 2005, from *http://home.twcny.rr.com/hiemstra/sdlindex.html*

Brookfield, S.D. (1986). *Understanding and facilitating adult learning*. San Francisco: Jossey-Bass.

Brown, R.E. (2001). The process of community building in distance learning classes. *Journal of Asynchronous Learning Environments, 5*(2). Retrieved August 9, 2002, from *www.aln.org/alnweb/journal/vol5_issue2/brown/brown.htm*

Center for Instructional Development and Research. (2003). Inclusive teaching. Retrieved February 26, 2004, from *http://depts.washington.edu/cidrweb/inclusive/*

Clark, R. (1994). Media will never influence learning. *Educational Technology Research and Development, 42*(2), 21-29.

Daughenbaugh, R., Ensminger, D., Frederick, L., Surry, D. (2002, April). Does personality type effect online versus in-class course satisfaction? *Teaching, Learning, & Technology: The Connected Classroom. Proceedings of the Annual Mid-South Instructional Technology Conference*, Murfreesboro, TN.

Dooley, K.E., Kelsey, K.D., & Lindner, J.R. (2003). Doc@Distance: Immersion in advanced study and inquiry. *Quarterly Review of Distance Education, 4*(1), 43–50.

Dooley, K.E., & Lindner, J.R. (2002). Competencies for the distance education professional: A self-assessment to document professional growth. *Journal of Agricultural Education, 43*(1), 24-35.

Fulford, C. P., & Zhang, S. (1993). Perceptions of interaction: The critical predictor in distance education. *The American Journal of Distance Education, 7*(3), 8-21.

Giguere, P., & Minotti, J. (2003). Developing high-quality Web-based training for adult learners. *Educational Technology, 43*(4), 57-58.

Grow, G.O. (1991). Teaching learners to be self-directed. *Adult Education Quarterly, 41*(3), 125-149. Expanded version available from *www.long leaf.net/ggrow*

Grow, G.O. (1994). In defense of the Staged Self-Directed Learning Model. *Adult Education Quarterly, 44,* 109-114.

Harter, S. (1978). *Effectance motivation reconsidered: Toward development model. Human Development, 21,* 34-64.

Herrnstein, R.J., & Murray, C. (1994). *The bell curve: Intelligence and class structure in American life*. New York: The Free Press.

Hillman, D.C., Willis, D.J., & Gunawardena, C.N. (1994). Learner-interface interaction in distance education: An extension of contemporary models and strategies for practitioners. *The American Journal of Distance Education, 8*(2), 30-42.

Howland, J.L., & Moore, J.L. (2002). Student perceptions as distance learners in Internet-based courses. *Distance Education, 23*(2), 183-195.

Kearsley, G., & Shneiderman, B. (1999). Engagement theory: A framework for technology-based teaching and learning. Retrieved August 8, 2002, from *http://home sprynet.com/~gkearsley/engage.htm*

Kinshuk, Y.A. (2003). Web-based asynchronous synchronous environment for online learning. *USDLA Journal, 17*(2), 5–18. Retrieved January 7, 2005, from *www.usdla.org/html/journal/ED_APR03.pdf*

Kirkup, G., & von Prummer, C. (1990). The needs of women distance education students. *Journal of Distance Education, 5*(2), 9-31.

Knowles, M.S. (1990). *The adult learner: A neglected species*. Houston, TX: Gulf.

Knowles, M.S., Holton, E.F., III, & Swanson, R.A. (1998). *The adult learner: The definitive classic in adult education and human resource development*. Woburn, MA: Butterworth-Heinemann.

Lindner, J.R., & Dooley, K.E. (2002). Agricultural education competencies and progress towards a doctoral degree. *Journal of Agricultural Education, 43*(1), 57-68.

Lindner, J.R., Dooley, K.E., & Hynes, J.W. (2003). Engagement and performance for female and male students in an online course. *The Texas Journal of Distance Learning, 1*(1). Retrieved February 9, 2004, from *www.tjdl.org/articles/1/engagement/*

Lindner, J.R., Dooley, K.E., & Murphy, T.H. (2001). Differences in competencies between doctoral students on-campus and at a distance. *American Journal of Distance Education, 15*(2), 25-40.

Lindner, J.R., Dooley, K.E., & Williams, J.R. (2003). Teaching, coaching, mentoring, facilitating, motivating, directing ... What is a teacher to do? *The Agricultural Education Magazine, 76*(2), 26-27.

Lindner, J.R., Hynes, J.W., Murphy, T.H., Dooley, K.E., & Buford, J.A., Jr. (2003). A comparison of on-campus and distance students progress through an asynchronously delivered Web-based course. *Journal of Southern Agricultural Education Research, 53*(1), 80-92. Retrieved February 9, 2004, from *http://pubs.aged.tamu.edu/jsaer/pdf/vol53/jsaer-53-080.pdf*

Lindner, J.R., & Murphy, T.H. (2001). Student perceptions of Webct in a Web supported instructional environment: Distance education technologies for the classroom. *Journal of Applied Communications, 85*(4), 36-47.

Lumsden, L.S. (1994). *Student motivation to learn* (Report No. 92). Eugene, OR: ERIC Clearinghouse on Education Management. (ERIC Document Reproduction Service No. ED370200).

Mandigo, J.L., & Holt, N.L. (2002). Putting theory into practice: Enhancing motivation through OPTIMAL strategies. *Avante, 8*(3), 21-29.

Merriam, S. (2001). Andragogy and self-directed learning: Pillars of adult learning theory. In S. Merriam (Ed.), *New Directions for Adult and Continuing Education, No. 89* (pp. 24-34). San Francisco: Jossey-Bass.

Merrow, J. (2003, February 5). Easy grading makes "deep learning" more important. *USA Today*, p. 12A.

Miller, G., & Pilcher, C.L. (2000). Are off-campus courses as academically rigorous as on-campus courses? *Journal of Agricultural Education, 41*(2), 65-72.

Miller, G., & Shih, C. (1999). A faculty assessment of the academic rigor of on and off-campus courses in agriculture. *Journal of Agricultural Education, 40*(1), 57-65.

Moore, M.G. (1989). Three types of interaction. *The American Journal of Distance Education, 3*(2), 1-7.

Murphy, T.H. (2000). An evaluation of a distance education course design for general soils. *Journal of Agricultural Education, 41*(3), 103-113.

Pappas, G., Lederman, E., & Broadbent, B. (2001). Monitoring student performance in online courses: New game–new rules. *Journal of Distance Education, 16*(2). Retrieved January 7, 2005 from *http://cade.icaap.org/vol16.2/pappasetal.html*

Rachal, J.R. (2002). Andragogy's detectives: A critique of the present and a proposal for the future. *Adult Education Quarterly, 52*(3), 210-227.

Rea, A., White, D., McHaney, R., & Sanchez, C. (2000). Pedagogical methodology in virtual course. In A. Aggarwal (Ed.), *Web-based learning and technologies: Opportunities and challenges* (pp. 135-154). Hershey, PA: Idea Group.

Russell, T.L. (1999). *The no significant difference phenomenon*. Raleigh: North Carolina State University, Office of Instructional Telecommunications.

Saba, F. (2000). Student attrition: How to keep your online learner focused. *Distance Education Report, 4*(14), 1-2.

Sansone, C., & Harackiewicz, J.M. (2000). *Intrinsic and extrinsic motivation: The search for optimal motivation and performance*. San Diego, CA: Academic Press.

Stokes, S.P. (2001). Satisfaction of college students with the digital learning environment. Do learners' temperaments make a difference? *The Internet and Higher Education, 4*(1), 31-44.

Taplin, M., Yum, J., Jegede, O., Fan, R., & Chan, M. (2001). Help-seeking strategies used by high-achieving and low-achieving distance education students. *Journal of Distance Education, 16*(1). Retrieved January 7, 2005 from *http://cade.icaap.org/vol16.2/pappasetal.html*

Tennant, M. (1992). The staged self-directed learning model. *Adult Education Quarterly, 42*, 164-166.

Weinstein, C.E. (2002). Learner control: The upside and the downside of online learning. *NISOD Innovation Abstracts, XXIV*(25), 1-2.

Wickersham, L.E, & Dooley, K.E. (2001). Attrition rate in a swine continuing education course delivered asynchronously. *Proceedings of the 28th Annual National Agricultural Education Research Conference, 48.*

Retrieved August 8, 2002, from *http://aaaeonline.ifas.ufl.edu/NAERC/2001/Papers/wickersh.pdf*

Zemke, R., & Zemke, S. (1981). 30 things we know for sure about adult learning. *Training Magazine, June,* 45-49

Section III

Systematic Instructional Design

Part III of the text focuses on the fundamental concepts of systematic instructional design. Writing instructional objectives, techniques for gaining attention and motivating learners, strategies to engage the learner actively, and methods to assess learning authentically are included. The focus is on student-centered, rather than teacher-centered design. We explore instructional design models and strategies that work well with the adult learner at a distance.

Chapter VI

Systematic Instructional Design

with
Atsusi Hirumi, University of Central Florida, USA

 Making Connections

In Part II, we explored adult learning principles, learner differences, and engaging learners to promote self-directed learning. Now, in Part III, we will examine systematic instructional design, including the student- or learner-centered approaches that promote lifelong learning. Although many trainers and instructors serve as both the content specialist and instructional designer, some institutions use a team approach with various people providing expertise. This chapter provides an overview of learner-centered instruction and instructional design models to help you or a team of developers conceptualize instructional planning. What are the components of instructional design? What is meant by teacher-centered versus learner-centered paradigms of instruction? How can we design instruction that will promote active learning and the use of critical and creative thinking skills?

Introduction

One of the most important considerations in the delivery of a program at a distance is the attention given to instructional design. "Instructional designs serve as mediators between the realms of learning theory and instructional practice, providing a means of developing interventions through which changes in learned capabilities can occur" (Wagner, 1994, pp. 20–21). Many instructional design models exist in the literature, but most models have common concepts. The Dick and Carey (1990) model is often used to design instruction with the process including developing broad instructional goals based on the needs of the audience, determining instructional objectives, developing means to determine if the objectives have been met, selecting strategies to implement the objectives, selecting media and methods for instruction, and providing formative and summative evaluation in an ongoing process.

Furthermore, instructional designers must also integrate learner-centered and self-directed approaches for distance learning. Learner-centered instruction considers a myriad of characteristics, processes, interactions, and delivery methods that result in effective teaching and learning. For example, asynchronous delivery strategies allow learners to complete work in their own time and location rather than be in the classroom at a specified time.

Related to learner-centered instructional design is the notion of self-directed learning. "As a person matures, his or her self-concept moves from that of a dependent personality toward one of a self-directing human being" (Merriam & Caffarella, 1999, p. 272). Adults prefer self-directed or self-designed activities more than relying on one medium for learning, and they also prefer control of the learning pace (Zemke & Zemke, 1984). Self-directed learning does not mean isolation. It may involve several resources, professionals, lectures, seminars, and face-to-face interactions. According to Grow (1991), adult learners progress from dependency to self-direction. "Some features of self-direction are distinctly situational: few learners are equally motivated toward all subjects. Some features appear to be deep, familial, perhaps even genetic traits of individual personalities—such as persistence" (p. 128).

Learner-Centered Instruction

Learner-centered instruction can be defined as any formal or nonformal education that accounts for a learner's cognitive and metacognitive factors, motivational and affective factors, developmental and social factors, and individual differences (APA, 1997). Cognitive and metacognitive factors include the nature of the learning process, goals of the learning process, construction of knowledge, strategic thinking, thinking about thinking, and context of learning. Motivational and affective factors include the emotional influence on learning, intrinsic motivation to learn, and effects of motivation on effort. Developmental and social factors include developmental and social influences on learning. Individual differences factors include differences in learning, diversity, standards, and assessment.

What are the hallmarks of learner-centered instruction? Learners in this setting are actively involved, applying knowledge to emerging issues, integrating discipline-based knowledge. Learners understand and value excellent work and become sophisticated knowers who are respected and valued. The design of learner-centered instruction requires a paradigm shift, challenging our beliefs about learning and the role that the instructor plays in the process (Huba & Freed, 2000). Figure 1 shows several of the key elements that embody the shift from teacher-centered to learner-centered instruction.

In a teacher-centered environment, the instructor's role often focuses on the transmission of knowledge. The learner-centered approach substitutes active learning experiences for passive activities such as lectures so that learners think about and solve problems requiring critical or creative thinking and includes the use of self-paced and/or cooperative learning (Felder & Brent, 1996). Re-

Figure 1. Comparison of teacher-centered and learner-centered instruction

PARADIGM SHIFT	
Teacher-centered instruction	**Learner-centered instruction**
• Knowledge transmitted	• Knowledge constructed
• Passive	• Active
• Context independent	• Context dependent
• Assessment separated	• Assessment integrated
• Competitive	• Cooperative

searchers have determined that the learner-centered approach may lead to "increased motivation to learn, greater retention of knowledge, deeper understanding, and more positive attitudes toward the subject being taught" (Felder & Brent, 1996, p. 43). As students experience learner-centered instruction, they may resist initially because of a mismatch between instructor role and learner readiness to perform in this setting. This resistance is part of the shift from teacher-centered instructional dependence to intellectual autonomy (Grow, 1991; Hirumi, 2002; Kloss, 1994). Instructors who are unaccustomed to a learner-centered approach may feel a bit uncomfortable in the beginning. The Thought and Reflection box below poses some possible concerns and potential solutions.

Thought and Reflection

A faculty member at a university who was designing her first course to deliver at a distance was encouraged by the instructional designer to consider using student-centered instructional design. The first question asked by the faculty member was, "Won't this active learning approach take too much time? How will I cover all my material?" The instructional designer assured the faculty member that much class time is wasted by assuming the learner is actually learning the material when simply copying notes from a lecture. Didactic materials can be provided through text-based or a brief video-based lecture (streaming video, Microsoft Producer file, or Camtasia file), but the essence of active learning is for the students to do more, and the instructor less. The faculty member then asked, "What about reading assignments? How can I make sure the learners understand the material independently?" The instructional designer suggested, "Provide links to readings with some advanced organizers to illustrate visually key points or you can prepare study guides that summarize critical questions in the readings. Even consider having the learners create the key-point summaries or process the material in a writing assignment." The faculty member replied, "This approach could really change the way I teach!" "Yes, that's the idea."

Learner-centered instruction should be clear and understandable, responsive to the ways in which students learn and communicate, and engaging, thereby incorporating the learners' interests and motivations (Egan & Gibb, 1997). In distance education, clarity should be achieved through organization and planning. By creating a detailed syllabus and interactive study guides, selecting course content, and selecting appropriate visualization tools, the instructor and instructional designer can provide a learning pathway that is easy to follow and understand.

To design effective learner-centered instruction, you must know your learners. By conducting an audience analysis to determine entry-level competence, motivations, and prior experience with technology, the instructor can develop instructional materials at the appropriate level and consider the use of individualized instructional sequences. For example, an instructor would need to design and deliver individualized instructional sequences to provide the greatest opportunity for learning when confronted with participants with dichotomous competencies; those who have little to no competence on any of the measurement items, and those who have high levels of competence on most items. Without a way of documenting learner competencies as they enter a program or course, an instructor cannot facilitate student-centered learning. As in most teacher-centered learning environments, they would be forced to teach or train to the average learner. Unfortunately, this is often the case, thereby providing course material that is too challenging for some learners and too simple for others.

Another factor that promotes learner-centered instruction is the incorporation of novel approaches to stimulate motivation and curiosity. In our own courses, we often use short video clips to gain the attention of the learner in a new way. These clips often include the use of props or analogies to connect the learners' prior knowledge with the material they will be learning. This also serves to build rapport and provide instructor immediacy (approachability) to help foster an interactive learning environment. The idea is a focus on *learning* rather than *teaching*.

 **Internet Connection**
www.aged.tamu.edu/classes/611

This link takes you to a graduate course, "Advanced Methods in Distance Education." You can view the instructional design features, including the use of video openers, as a novel approach to gain attention and stimulate motivation.

As noted earlier in this chapter, a move from teacher-centered to learner-centered instruction represents a major paradigm shift. To illustrate further this point, consider Figure 2. A teacher-centered approach assumes that teachers direct the learning process and control students' access to and application of information. Students are treated like empty vessels and learning is viewed as additive. Instruction is geared to the average students with everyone progressing at the same pace. Family and community members may influence student learning, but typically their activities (and resulting outcomes) are not planned as an integral part of the formal learning process.

Students, however, are not empty vessels and learning is an active, dynamic process. In student-centered environments, learners are given direct access to the knowledge base and work individually and in small groups to solve authentic problems. In such environments, parents and community members also have direct access to teachers and the knowledge base, playing an integral role in the process (Hirumi, 2002).

In short, the role of the instructor in a learner-centered environment is to facilitate learning. This includes providing resources to help the learner develop his or her skills in critical thinking, problem solving, and decision making by helping him or her access, interpret, organize, and otherwise apply information.

Figure 2. A comparison of teacher- and student-centered learning environments (Hirumi, 2002) (used with permission from the Association for the Advancement of Computing in Education)

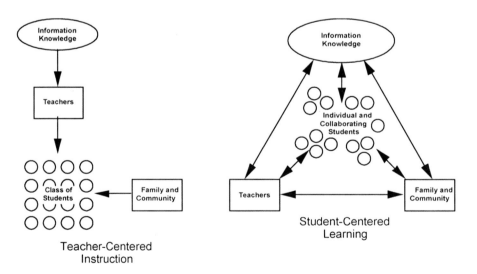

Table 1 further details the differences between teacher- and student-centered learning environments. After examining the contents of the table, ask yourself, "How do I describe in one sentence the actions that make up the fundamental differences between the teacher-centered and the student-centered learning environments?"

Now that we have defined and described learner-centered instruction, we will review instructional design models from two varying perspectives.

Table 1. A comparison of key instructional variables in teacher- and student-centered learning environments (Hirumi, 2002) (used with permission from the Association for the Advancement of Computing in Education)

Instructional Variables	Instructional Approach	
	Teacher Centered	Student Centered
Learning Outcomes	• Discipline-specific verbal information • Lower-order thinking skills (e.g., recall, identify, define) • Memorization of abstract and isolated facts, figures, and formulas	• Interdisciplinary information and knowledge • Higher-order thinking skills (e.g., problem solving) • Information processing skills (e.g., search for, access, organize, interpret, communicate information)
Goals and Objectives	• Teacher prescribes learning goals and objectives based on prior experiences, past practices, and state and/or locally mandated standards	• Students work with teachers to select learning goals and objectives based on authentic problems and students' prior knowledge, interests, and experience
Instructional Strategy	• Instructional strategy prescribed by teacher • Group paced, designed for average student • Information organized and presented primarily by teacher (e.g., lectures) with some supplemental reading assignments	• Teacher works with students to determine learning strategy • Self-paced, designed to meet needs of individual student • Student given direct access to multiple sources of information (e.g., books, online databases, community members)
Assessment	• Assessments used to sort students • Paper-and-pencil exams used to assess students' acquisition of information • Teacher sets performance criteria for students • Students left to find out what teacher wants	• Assessment integral part of learning • Performance based, used to assess students' ability to apply knowledge • Students work with teachers to define performance criteria • Students develop self-assessment and peer assessment skills
Teacher's Role	• Teacher organizes and presents information to group of students • Teacher acts as gatekeeper of knowledge, controlling students' access to information • Teacher directs learning	• Teacher provides multiple means for accessing information • Teacher acts as facilitator, helps students access and process information • Teacher facilitates learning
Students' Role	• Students expect teachers to teach them what's required to pass the test • Passive recipients of information • Reconstructs knowledge and information	• Students take responsibility for learning • Active knowledge seekers • Constructs knowledge and meaning
Environment	• Students sit individually in rows, information presented primarily via lectures and reading assignments	• Students work at stations, individually and in small groups, with access to electronic resources

Instructional Design Models

Instructional design is a systematic process or organized procedure for developing instructional materials (learning objects), including steps of analysis (defining what is to be learned), designing (specifying how the learning should occur), developing (producing the learning objects), implementing (using delivery strategies), and evaluating (determining if the objectives have been satisfied). Instructional design emerges from system theory. The main purpose is to construct effective delivery of a learning experience (Morrison, Ross, & Kemp, 2004).

We introduced the concept of learner-centered instruction in the previous section. Some content areas and potential audiences may still rely on teacher-directed instruction. If this is the case, you may need to use Grow's model (discussed in chapter V) to help learners gradually become more self-directed.

We will discuss two major areas of learning theory: behaviorist (introduced in chapter III) and constructivist. Remember that your own personal philosophy and type of audience and content will determine the approaches that you may choose to use. If you would like a detailed description of these paradigms of instruction, see the Internet Connection section below.

 Internet Connection
http://carbon.cudenver.edu/~mryder/itc_data/idmodels.html

This site provides a detailed listing of links on instructional design to aid in understanding the theoretical components of behaviorism (prescriptive models) and constructivism (phenomenological models).

Behaviorist Learning Theory

Behavioral theorists are interested in observable actions that will "change" the learner. Reinforcement of desired responses is a critical component. Educational goals are operationally defined, empirically measurable, observable behaviors. "[L]earning is achieved through frequent response and immediate reinforcement of appropriate behavior" (Villaba & Romiszowski, 2001, p.

327). Hence, behaviorist principles provided the foundation for the antecedents of Web-based training: drill and practice, sound motion pictures, silent filmstrips, and instructor manuals. It also served as a foundation for task analysis concepts, classroom management techniques, commercially marketed programmed instructional modules, and development of the early computer-assisted instruction models (Hilgard & Bower, 1975; Tennyson, Schott, Seel, & Dijkstra, 1997).

Smith and Ragan (2000) emphasize that behaviorist principles allowed for objective research in education by eliminating the "form of encounter" variable, allowing for valid and reliable empirical evidence to be acquired in areas such as practice, feedback, sequence, criterion referenced assessment, small frames of instruction, and self-pacing. These attributes are still considered valuable in Web-based training and form the basis for popular instructional design modules, such as the one posited by Smith and Ragan. Smith and Ragan's instructional design model consists of three main stages: analysis (learning environment, learners, learning task, write test items), strategy (determine organizational strategies, delivery strategies, management strategies, and write and produce instruction), and evaluation (conduct formative evaluation and revise previous steps). The specific tasks associated with each stage, in turn, are grounded in research and theory.

Here are a few other examples. Dick and Carey developed an instructional design model that was first published in 1978 and is still considered a seminal piece by trainers and instructional designers in military and corporate settings. This model describes the instructional design process from stating goals and writing objectives to developing materials, assessing instruction, and grading. It focuses on the development of criterion-referenced test items to determine if instructional objectives have been met and formative evaluation to ensure the effectiveness and efficiency of the instruction (Dick & Carey, 1990; Dick, Carey, & Carey, 2001).

Another common instructional design model is called "ADDIE": **A**nalyze needs, **D**esign instruction, **D**evelop materials, **I**mplement activities, and **E**valuate participant progress and instructional effectiveness (Hall, 1997; Powers, 1997). An audience analysis, budget, and due dates are some of the considerations during the analysis phase. Selecting the most appropriate environment, writing instructional objectives, selecting the overall approach and program look and feel, and designing the course content are components of the design phase (Driscoll, 1998; Porter, 1997). In the development phase, media are obtained and/or created and instructional designers determine appropriate interactions to encourage learners to construct a supportive social environment (Porter, 1997; Simonson, Smaldino, Albright, & Zvacek, 2003). In the implementation phase, all materials are available and maintained on the course Web site. Preparation for technical difficulties and alternative plans are also communicated (Simonson et al., 2003). Finally, the designer creates appropriate assessment and evaluation procedures, whether those are criterion-referenced test items, research papers, class participation, or competency skills (Powers, 1997). Instructional designers also plan formative evaluations to test for problems or misunderstanding (Schrum, 1998) as well as summative evaluation to determine the overall effect of the course (Bourne, McMaster, Rieger, & Campbell, 1997).

This section represents only a few instructional design models in the literature. Although semantics vary, the overall concepts/stages are fairly consistent. Each has an analysis stage, generally followed by writing instructional objectives, deciding upon the techniques to use to develop and deliver instruction, and then methods to evaluate the results to see if the objectives have been met.

However, some argue that traditional instructional design models grounded in behavioralistic theories do not account for the dynamic nature of human learning (Halff, 1988), and thus, are unsuitable for facilitating learner-centered instruction. Instructional designers applying traditional design models are accustomed to generating prepackaged instructional sequences that a trainer/instructor can pick up and replicate in traditional classroom settings. However, prepackaged instruction that sets goals, strategies, and assessments prior to delivery and remain set during the learning process may not be appropriate with adult learners, particularly at a distance. According to adult learning theory, every individual derives meaning from content in a unique way, based upon their previous experiences. Thus, instructional goals and activities cannot necessarily be prescribed or prepackaged. Instructional methods and materials must be dynamic in learner-centered environments to meet individual needs and interests, which leads us (and others) to look at design from another viewpoint.

Constructivist Learning Theory

In contrast to the behaviorists, the rationalist epistemological movement in Germany in the 1920s gave rise to Gestalt theory. The interest was in perception and problem-solving processes, attributing the locus of control for a learning activity to the learner, in contrast to the behaviorist belief that it lay with the environment (Hilgard & Bower, 1975; Merriam & Caffarella, 2000; Tennyson, Schott, Seel, & Dijkstra, 1997). "New information is built on existing knowledge structures or schemas, relevant processing activities are stimulated, and specific strategies are taught to ensure that the learner efficiently acquires the information or solves the problems" (Villaba & Romiszowski, 2001, p. 327).

Many educational psychologists find that behaviorist approaches do not fully explain how and why people learn and do not meet their needs. In areas of critical and creative thinking, learning strategies may be more concerned with what is unobservable, what is going on *in* the mind (or the black box as previously mentioned in chapter III). These theories are based on the work of John Dewey, Lev Vygotsky, Jean Piaget, and Jerome Bruner. "Learning from this perspective [constructivist] is viewed as a self-regulatory process of struggling with the conflict between existing personal models of the world and discrepant new insights, constructing new representations and models of reality as a human meaning—making venture with culturally developed tools and symbols, and further negotiating such meaning through cooperative social activity, discourse, and debate" (Fosnot, 1996, p. ix). In a constructivist environment, the learner has a multitude of ways in which to address an issue, such as experimentation, computer models, and dialogue with peers. Learners are active, independent participants. Cognitive experiences situated in authentic activities, such as project-based learning, cognitive apprenticeships, or case-based learning environments, result in richer and more meaningful learning experiences. Social negotiation allows learners to process and test their construction through discourse, dialogue, and collaboration.

For a wonderful comparison of behaviorist and constructivist approaches, see the Internet Connection below.

 Internet Connection

http://depts.washington.edu/eproject/Instructional_Design_Approaches.htm

This site provides a comparison of the two instructional design approaches by learning theorists, learning outcomes, instructor role, student role, activities, and assessment.

Instructional design models based solely on constructivist theory are difficult to find. Constructivist design principles have been published (e.g., Hannifin, Hannifin, Land, & Oliver, 1997; Honebein, 1996; Jonassen, 1999), but they are difficult to implement because they present educators with heuristics (general principles) rather than algorithms (step-by-step processes). To help educators apply constructivist and learner-centered approaches to teaching and learning, Hirumi (2002) developed the Student-Centered, Technology-Rich Learning Environment (SCenTRLE) model.

Multiple perspectives, active participation, the construction of meaning, argumentation, and reflection are all key components of SCenTRLE. SCenTRLE presents eight instructional events or stages for facilitating construction of knowledge and the development of metacognitive skills associated with lifelong learning:

- *Event 1 – Set Learning Challenge.* The challenge may take the form of an instructional goal, goal statements, or learning outcome. The challenge should situate learning within an authentic context and describe what the learners should be able to do as a result of the instruction.

- *Event 2 – Negotiate Learning Goals and Objectives.* At this stage, the learners should assess their learning requirements and work with the instructor to set individual learning goals and objectives. This negotiation can occur through discussion, learner assessments, preliminary definition of goals and objectives, feedback from the instructor, revision (if necessary), and continuous monitoring.

- *Event 3 – Negotiate Learning Strategy.* During this event, learners determine how they will achieve their learning goals and objectives. Learners and the instructor discuss various methods (individually or in small or large groups) and determine jointly the best course of action. Learners begin developing an important skill associated with independent learners during this process, namely, being able to discern what strategies or learning resources are most effective and efficient for themselves.

- *Event 4 – Construct Knowledge.* This event requires learners to work individually and in groups to derive meaning from various resources identified during Event 3 and to construct their skills and knowledge. The instructor monitors group and individual progress, answers questions, and facilitates learning when necessary.

- *Event 5 – Negotiate Performance Criteria.* A significant difference between a novice and an expert is that an expert can judge the quality of his or her own work as well as the work of others; a novice cannot do so. During Event 5, learners work with the instructor to define performance criteria for their selected goals and objectives (a characteristic of self-directed, lifelong learners). Emphasis is on the assessment of work samples in a portfolio with specificity included on the characteristics of excellence.

- *Event 6 - Conduct Self, Peer, and Expert Assessments.* After establishing performance criteria and generating work samples, the learners are asked to assess themselves and have at least one other adult also assess their work using the performance criteria and assessment rubrics generated in Event 5. The key is to obtain feedback for continuous improvement.

- *Event 7 – Monitor Performance and Provide Feedback.* This event takes place throughout the learning process. The instructor monitors work, examines documents, replies to e-mail, asks the learner how he or she is doing, and so forth. Learners also provide feedback to each other.

- *Event 8 – Communicate Results.* Informally, communications are used for self, peer, and expert assessments to generate feedback. Learners are also asked to prepare, present, and submit a portfolio with assessment rubrics, work samples, and a narrative description. The production and selection of work samples illustrate achievement of their goals and objectives, and the narrative provides an opportunity for reflection and articulation of what they have learned.

The eight events in the SCenTRLE model provide a framework appropriate for learner-centered instructional design and assessment. We do not argue that systematic design models, such as the one posited by Dick and Carey (1990), are unsuitable for facilitating learner-centered instruction. Instructional design models should in fact be used to design learner-centered instruction, but the designer should consider the purpose of the various steps posited by each model. For example, in the Dick and Carey model, educators and instructional designers are directed to conduct learner, task, content, subject matter, and context analysis to define and prescribe learning objectives. In the SCenTRLE model, designers are urged to conduct analyses, not to prescribe objectives, but to identify objectives to be used later by the instructor as a foundation for negotiating learning goals and objectives by the learner (Event 2). The Dick and Cary model directs designers to develop and prescribe instructional strategies for facilitating learner achievement of defined objectives. In SCenTRLE, designers identify strategies, but rather than prescribe, they identify them for later use for negotiation by the learner (Event 3). The Dick and Carey model also presents steps for establishing and prescribing performance criteria. SCenTRLE recommends similar techniques, once again for negotiation rather than prescription. In short, the SCenTRLE model has instructors apply an instructional design model so that they can help learners define their own learning goals and objectives, strategies, and performance criteria.

Concept Mapping

One exercise that is helpful for instructional designers is to create a concept map of the systematic instructional design process to help visualize the necessary components. Allow us to operationalize a few terms. A concept is a mental image, key idea, or term (such as events, objects, values, skills, or attitudes). A proposition is two or more concepts linked appropriately by linking words (such as are, where, the, in, how, does, makes, etc.). A concept map is a schematic device for representing a set of concept meanings embedded in a framework of propositions. It is a drawing in which key ideas are identified and the relationship between and among them is described.

The strategy for using concept maps includes the use of circles or squares placed around statements or descriptions of concepts. Concepts may be placed on the page in a hierarchical form from global to specific, from complex to easy, from superordinate to subordinate, or by some other meaningful pattern. One set of concepts may be linked not only among themselves but also to other sets of concepts. If relationships (valid, significant, and/or creative) are found, they can be diagrammed and linked.

The concept map serves as a schema or a mental map that helps to determine where we are and where we are going. Similarly, instructors can use concept maps as an instructional technique to help learners create a mental map or schema to connect concepts to other related facts. Concept maps serve as organization tools. Concept maps show links and associations while maintaining the overall picture. Concept maps can be used to help learners explore what they already know, serve as a learning route, help learners extract meaning from texts, laboratory, studio, and/or field studies, or help learners plan a pathway for writing a paper or completing an assignment. To practice the use of a concept map, we will ask you to use this technique in the application exercise for this chapter. We have included as Figure 3 below a creative example called

Figure 3. Instructional design concept map example

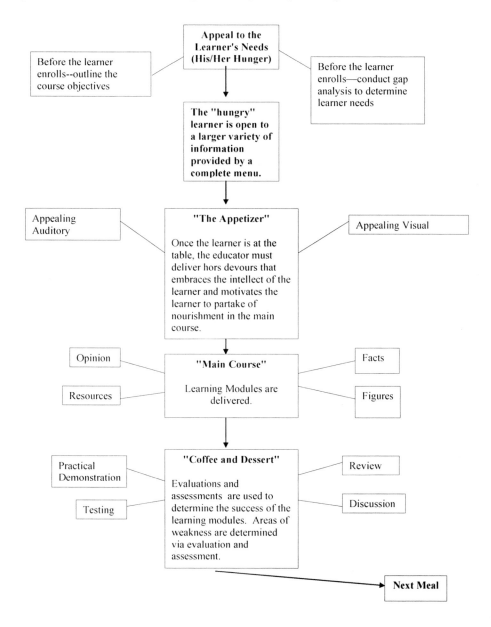

"The Complete Meal for the Mind," prepared by a graduate student, James Michael Farrow (used with permission).

Conclusion

In this chapter, we have explored instructional design from two varying perspectives and the development of learner-centered instructional strategies. The classic instructional systems design model is an outcomes-based model with behavioral and cognitive approaches to instruction. It is described as a model that stresses performance based upon the skill level of the learners with opportunities for revision to promote instruction and learning (Dick, Carey, & Carey, 2001). We share another possibility, SCenTRLE, for a constructivist approach to instructional design. This model promotes a learner-centered focus to promote active engagement and critical/creative thinking. In the next chapter, we will discuss Gagné's Nine Events of Instruction and how to gain attention and stimulate motivation as an important dimension of an active distance education environment.

 Application Exercise

There are a variety of instructional design models. Although they use different terms, they basically have similar components. Create a concept map of what you think are the critical components of systematic instructional design (the big picture—not specifics) with relationships indicated by arrows or another form of "connections." Be creative!

References

American Psychological Association (APA). (1997). *Learner-centered psychological principles: A framework for school redesign and reform.* Retrieved October 9, 2003, from *www.apa.org/ed/lcp.html*

Bourne, J.R., McMaster, E., Rieger, J., & Campbell, J.O. (1997, August). Paradigms for online learning. A case study in the design and implementation of an asynchronous learning networks (ALN) course. *Journal of Asynchronous Learning Networks, 1*(2). Retrieved January 8, 2005 from *http://www.sloan-c.org/publications/jaln/v1n2/v1n2_bourne.asp*

Dick, W., & Carey, L. (1990). *The systematic design of instruction.* Glenview, IL: Scott Foresman.

Dick, W., Carey, L, & Carey, J.O. (2001). *The systematic design of instruction* (5th ed.). New York: Addison-Wesley.

Driscoll, M. (1998). *Web-based training.* San Francisco: Jossey-Bass/Pfeiffer.

Egan, M.W., & Gibb, G.S. (1997). Student-centered instruction for the design of telecourse. In T.E. Cyrs (Ed.), *Teaching and learning at a distance: What it takes to effectively design, deliver, and evaluate program* (pp. 33-39). San Francisco: Jossey-Bass.

Felder, R.M., & Brent, R. (1996). Navigating the bumpy road to student-centered instruction. *College Teaching, 44*, 43-47.

Fosnot, C.T. (Ed.). (1996). *Constructivism: Theory, perspective, and practice.* New York: Teachers College Press.

Grow, G.O. (1991). Teaching learners to be self-directed. *Adult Education Quarterly, 41*(3), 125-149.

Halff, H.M. (1988). Curriculum and instruction in automated tutors. In M.C. Polson, & J.J. Richardson (Eds.), *Foundations of intelligent tutoring systems* (pp. 79-108). Hillsdale, NJ: Lawrence Erlbaum.

Hall, B. (1997). *Web-based training cookbook.* New York: John Wiley & Sons.

Hannafin, M.J., Hannafin, K.M., Land, S.M., & Oliver, K. (1997). Grounded practice and the design of constructivist learning environments. *Educational Technology Research and Development, 45*(3), 101-117.

Hilgard, E.R., & Bower, G.H. (1975). *Theories of learning.* Englewood Cliffs, NJ: Prentice Hall.

Hirumi, A. (2002). Student-centered, technology-rich, learning environments (SCenTRLE): Operationalizing constructivist approaches to teaching and learning. *Journal for Technology and Teacher Education, 10*(4), 497-537.

Honebein, P.C. (1996). Seven goals for the design of constructivist learning environments. In B. Wilson (Ed.), *Constructivist learning environments: Case studies in instructional design* (pp. 3-8). Englewood Cliffs, NJ: Educational Technology Publications.

Huba, M.E., & Freed, J.E. (2000). *Learner-centered assessment on college campuses: Shifting the focus from teaching to learning.* Needham Heights, MA: Allyn & Bacon.

Jonassen, D. (1999). Designing constructivist learning environments. In C.M. Reigeluth (Ed.), *Instructional-design theories and models: A new paradigm of instructional theory* (Vol. II, pp. 215-239). Mahwah, NJ: Lawrence Erlbaum.

Kloss, R.J. (1994). A nudge is best: Helping students through the Perry scheme of intellectual development. *College Teaching, 42*(4), 151-158.

Merriam, S.B., & Caffarella, R.S. (1999). *Learning in adulthood: A comprehensive guide.* San Francisco: Jossey-Bass.

Morrison, G.R., Ross, S.M., & Kemp, J.E. (2004). *Designing effective instruction* (4th ed.). Hoboken, NJ: John Wiley & Sons.

Porter, L.R. (1997). Creating the virtual classroom, learning with the Internet. New York: John Wiley & Sons.

Powers, S.M. (1997). *Designing an interactive course for the Internet. Contemporary Education, 68,* 194-196.

Schrum, L. (1998). On-line education: A study of emerging pedagogy. *New Directions for Adult and Continuing Education, 78,* 53-61.

Simonson, M., Smaldino, S., Albright, M., & Zvacek, S., (2003). *Teaching and learning at a distance: Foundations of distance education.* Upper Saddle River, NJ: Merrill Prentice Hall.

Smith, P.L., & Ragan, T.J. (2000). The impact of R.M. Gagné's work on instructional theory. In R.C. Richey (Ed.), *The legacy of Robert M. Gagné* (pp. 147–181). Syracuse, NY: ERIC Clearinghouse on Information & Technology.

Tennyson, R.D., Schott, F., Seel, N., & Dijkstra, S. (Eds.). (1997). *Instructional international perspective design.* Mahwah, NJ: Lawrence Erlbaum.

Villaba, C., & Romiszowski, A.J. (2001). Current and ideal practices in designing, developing, and delivering Web-based training. In B. Kahn (Ed.), *Web-based training* (pp. 325-342). Englewood Cliffs, NJ: Educational Technology Publications.

Wagner, E.D. (1994). In support of a functional definition of interaction. *The American Journal of Distance Education, 8*(2), 6-29.

Zemke, R., & Zemke, S. (1984). Thirty things we know for sure about adult learning. *Innovation Abstracts, 6*(8). Retrieved January 7, 2005, from *www.hcc.hawaii.edu/intranet/committees/FacDevCom/guidebk/teachtip/adults-3.htm*

Chapter VII

Writing Instructional Objectives

 **Making Connections**

In the last chapter we discussed learner-centered instruction and gave you an overview of systematic instructional design. One of the first considerations after determining the needs of your audience, the potential learners, and the content to be delivered is to formulate instructional objectives. Instructional objectives are written by the instructor to guide the design process, and must consider distance education delivery strategies and principles of adult learning. Often these objectives will be negotiated with the learner so that they will meet their individual needs (e.g., learning contracts). Keeping in mind that learners have diverse learning needs and preferences, it is important to understand the three major domains of learning: cognitive, affective, and psychomotor. Helping you to do so are guideposts to ensure that the instructional objectives are written so that they measure the intended outcomes. How do you write instructional objectives that are specific and measurable? Why is this important?

Introduction

Effective instruction begins with the establishment of instructional goals and objectives (Brahier, 2000). Goals describe learner outcomes expected upon completion of a course or instructional unit. Instructional goals should be general, observable, and challenging. *To develop a greater appreciation for using geometry to solve real-word problems* is an example of an instructional goal. Instructional goals should be directly related to the content being taught and the competencies being developed (Newcomb, McKracken, Warmbrod, & Whittington, 2004).

Many instructors confuse the terms "instructional goals" and "objectives," believing them to be synonymous. They are not. An instructional objective is a statement describing a proposed "change" of what the learner can do when he or she has successfully completed a learning experience. Objectives should be **S**pecific, **M**easurable, **A**ttainable, **R**elevant, and **T**imed (SMART). An objective is specific or precise so that the instructor and learner can determine whether the objective has been met. If the objective is measurable, the instructor should be able to observe the action or change and thus provide feedback for improvement if needed (Mager, 1997). When writing objectives, instructors make decisions about the content and establish parameters to help define and limit the content (Newcomb et al., 2004). For example, *Given a scalene triangle, students will be able to prove that the sum of the measures of the angles of a triangle is 180* is an example of an instructional objective. Note that the three characteristics of a good, defensible behavioral instructional objective are present in the statement of this objective. First, the terminal behavior expected and accepted as evidence is identified by name, namely, "students will be able to prove that the sum of the measures of the angles of a triangle is 180." Second, the important conditions under which the behavior will be expected to occur are described to define further the desired behavior, namely, "Given a scalene triangle." Third, a criterion of acceptable performance is included or implied that describes how well the student must perform to be considered acceptable, namely, "students will be able to prove" (student either can prove or cannot prove). Goals are general; objectives are specific.

Domains of Learning

Instructional objectives can be classified into three domains of learning: cognitive (Bloom & Krathwohl, 1956), affective (Krathwohl, Bloom, & Masia, 1964), or psychomotor (Harrow, 1972). Instructional objectives are written for the cognitive domain when there is a lack of knowledge. Cognitive objectives are for mental skills and abilities; the acquisition of information and concepts related to the course content. The affective domain considers feelings and emotions (attitudinal change) as a result of the educational experience. Affective objectives are written when there is a lack of desire to help foster certain values, attitudes, and preferences. The psychomotor domain requires competence in the performance of procedures, operations, methods, and techniques (skills). Psychomotor objectives include physical movement, coordination, and use of motor-skill areas such as being proficient on the computer keyboard. Development of these skills requires practice and is often measured in terms of speed, precision, or distance. Objectives are written with a specific domain in mind, but obviously there are overlaps because we cannot isolate knowledge, attitudes, and skills as information is processed in the brain.

Each of these domains includes a hierarchy, or levels, from simplest to complex. We will describe these levels for each domain with examples of observable action verbs that serve as the catalyst for writing objectives and for assessing whether the objective has been achieved.

Cognitive Domain

When first writing instructional objectives, it is important to determine the primary domain that describes what the learner will do as a result of the instruction. If the learner lacks knowledge, then the cognitive domain will be your choice. Then, it is imperative to think about how the learner will do this. How difficult is the concept? Does it require knowledge related to previous instruction that would require scaffolding or applying other concepts? Does it require higher order, critical, or creative thinking?

We find it useful to use the taxonomy created by Bloom and Krathwohl (1956) in order to choose the level and verbs that best describe what we want learners to be able to do (Table 1). *Knowledge* is the simplest learning process and *evaluation* the most complex. In other words, a learner must have some basic

Table 1. Cognitive domain levels and verbs

KNOWLEDGE Learn specific facts, ideas, and vocabulary and reiterate in similar form	COMPREHENSION Communicate knowledge and interpret previous learning	APPLICATION Use learned knowledge and interpret previous situation	ANALYSIS Break down an idea into its parts and perceive the interrelation ship	SYNTHESIS Use elements in new patterns and relationships	EVALUATION Make decisions or judgments based on chosen criteria of standards
choose	calculate	adapt	analyze	alter	accept
collect	categorize	apply	appraise	change	appraise
copy	change	assemble	arrange	combine	argue
define	classify	build	break down	compose	assess
describe	communicate	calculate	categorize	construct	challenge
discover	define	compute	classify	create	choose
experiment	describe	construct	compare	depict	classify
find	discuss	demonstrate	contrast	design	criticize
identify	distinguish	develop	decipher	develop	critique
indicate	expand	discover	deduce	devise	debate
label	generalize	discuss	determine	estimate	decide
list	illustrate	dramatize	diagram	expand	defend position
locate	in your own words	draw	differentiate	forecast	determine
mark	infer	experiment	causes	form a new	discuss
match	inform	formulate	dissect	generate	document
name	interpret	gather	distinguish	hypothesize	draw conclusions
narrate	name alternatives	illustrate	examine	imagine	editorialize
observe	outline	make	experiment	incorporate	establish
produce	paraphrase	make use of	explain	infer	evaluate
provide	rearrange	manipulate	generalize	integrate	hypothesize
read	reconstruct	operate	group	invent	interpret
recall	relate	organize	illustrate	modify	judge
recite	restate	practice	infer	organize	justify
recognize	retell	prepare	investigate	plan	prioritize
record	reword	put into action	model	predict	rank
relate	rewrite	relate	modify	produce	rate
repeat	select	report use	order	relate	recommend
report	summarize	revise	outline	simplify	refute
select	tell the meaning of	search	predict	synthesize	reject
sort	translate	show	question		speculate
spell	verbalize	solve	reason why		test
state	write	survey	relate		value
tell		tell	separate		verify
underline		try	sequence		
write		utilize	solve		
			specify		
			subdivide		
Source: Bloom & Krathwohl (1956)					

understanding before he or she can process more complex information. Several action verbs are repeated in more than one level, but the context or description surrounding the objective statement should clarify the level to which it belongs. We will describe this in the objective writing section later. The first three levels (knowledge, comprehension, and application) are considered lower-level thinking, and the last three (analysis, synthesis, and evaluation) are higher-order thinking, sometimes referred to as HOTS (higher-order thinking skills).

Objectives are more easily written for lower-order cognitive skills. Instructional objectives should be written in such a manner that all levels of knowledge are addressed. Instructors often write objectives at lower cognitive levels, yet they test or assess learning at a higher level. It is important to match objectives and assessment. If an objective is written correctly, the instructional strategy and assessment of learning outcomes will already be determined. A little extra work at this step will save time later in the design and delivery stages of instruction. We will now move to the affective domain.

Affective Domain

The affective domain presents more challenges in terms of writing objectives and measuring outcomes than do the other domains. On the surface, it seems to be an easy domain with which to work. For example, just by having learners attend and participate in a teaching–learning activity, you are working within this domain of learning. However, the domain really is complex, especially when attempting to evaluate outcomes. For example, how do you determine if there is a desirable change in feelings or attitudes? The five levels in the affective domain range in complexity from simply receiving phenomena to characterizing and internalizing values. You will note in Table 2 that the verbs listed under each

Table 2. Affective domain levels and verbs

RECEIVING Willingness to attend; awareness	RESPONDING Active participation; attends and reacts	VALUING Worth or value the learner attaches to an object or phenomenon, from simple acceptance to commitment	ORGANIZING Building a consistent value system, resolving conflicts; prioritizing values	CHARACTERIZING Internalize values; value system is pervasive, consistent, predictable, and characteristic of the learner; lifestyle
ask	answer	complete	adhere	act
choose	assist	demonstrate	alter	discriminate
describe	aid	differentiate	arrange	display
follow	discuss	explain	combine	influence
give	greet	form	compare	listen
hold	help	initiate	defend	perform
identify	label	invite	explain	practice
locate	perform	join	generalize	qualify
name	practice	justify	integrate	question
select	present	propose	modify	revise
use	read	report	organize	serve
view	select	share	relate	solve
watch	write	work	synthesize	verify
Source: Krathwohl, Bloom, & Masia (1964)				

level describe actions for that particular level and serve to differentiate each level of the domain from the other domains. You will also note that much subjectivity is involved when one attempts to determine if the desired outcome has occurred.

Once a learner moves into the higher levels, he or she is beginning to internalize a set of specified values, expressed by the learner's overt behavior. Once again, the sample action verbs are repeated in different levels and are also present in the cognitive domain. The difference is in the context of the learning environment—the audience, the content, and the primary domain where the learner and instructor would like to see change or improvement as a result of instruction.

Affective domain objectives are often difficult to measure. Mager (1997) cautions that if the objective does not include a performance, then it is not an objective. Statements about the affective domain are often statements of inference, not performance. They may be predictions about future behavior that can be determined through evidence of what people say and do. Therefore, instructors should develop affective objectives that will satisfy the intent through a description or observation of the performance (Mager, 1997).

Psychomotor Domain

When we think about the psychomotor domain, we think of talented athletes or performers such as Mikhail Baryshnikov. If you have seen Baryshnikov perform, then you can imagine the fluid motion. That is the highest level of psychomotor performance. As instructors in distance education, the skills we teach may not require origination to that extent, but some of the work in virtual reality and simulation do require significant psychomotor skill to accomplish a task.

There are eight levels in the psychomotor domain (Table 3). Notice that the action verbs provided for mechanism and complex overt response are the same. Instructional objectives include adverbs or adjectives that will indicate that the performance is quicker, better, more accurate, and so forth. Think of the psychomotor domain in terms of what a person can *do*. Using a keyboard, calibrating an instrument, and adjusting a thermostat are some examples. However, the objective is not just to use the keyboard or to calibrate an instrument. We must remember (1) that a level of terminal behavior is expected and accepted as evidence, (2) the conditions under which that desired or expected behavior will be expected to occur must be defined, and (3) criteria of acceptable performance must be established by describing how well the learner must perform in order for that performance to be considered acceptable. For example, the instructional objective, *Be able to weld ¼-inch mild steel plate together four times consecutively using 5/32-inch diameter E-6011 electrodes to make single vee butt welds that can be bent double without the weld breaking* meets the three criteria of the psychomotor domain.

In distance education we struggle with teaching "hands-on" skills that are normally taught in a laboratory setting. As instructors and instructional designers, we must remember that many learners need the tactile or psychomotor component to be successful. For example, asking learners to type their responses engages three different parts of the brain and all three domains of learning.

Table 3. Psychomotor domain levels and verbs

PERCEPTION	SET	GUIDED RESPONSE	MECHANISM	COMPLEX OVERT RESPONSE	ADAPTION	ORIGINATION
Sense organs guide motor activity	Readiness to take action	Imitation; trial and error	Do alone in less time without describing the steps; responses become habitual; move with some confidence and proficiency	Do without error; skillful performance of motor acts that involve complex movement patterns; performing without hesitation; quick, accurate, and highly coordinated performance	Do in a different way; skills are well developed and can be modified to fit special requirements	Do in a new way; create new movement pattern to fit a particular situation or problem; highly developed skills
choose	begin	copy	assemble	assemble	adapt	arrange
describe	display	trace	calibrate	calibrate	alter	build
detect	explain	follow	construct	construct	change	combine
differentiate	move	react	dismantle	dismantle	rearrange	compose
draw	proceed	reproduce	display	display	reorganize	construct
feel	react	respond	fasten	fasten	revise	create
identify	show	watch	fix	fix	vary	design
isolate	state		grind	grind		initiate
relate	volunteer		heat	heat		make
select			manipulate	manipulate		originate
			measure	measure		
			mix	mix		
			sketch	sketch		

Source: Harrow (1972)

For instructional objectives to be effective, they must be clearly written and provided to the learners. Subsequent instructional sequences (events) and activities should focus on helping learners achieve the stated objectives. Instruction and assessment of learner outcomes should also focus on helping learners achieve the goals and objectives of the course or program. Objectives are useful for providing a sound basis for selecting instructional materials and delivery strategies, developing and negotiating measurable results and outcomes, and for communicating the focus of the course or program to the learner (Mager, 1997). A good instructional objective is functional, definite, attainable, measured, and accepted by the learners.

Teaching Laboratory/ Hands-On Skills at a Distance

One of our Doc@Distance students, Kim Hays, is studying teaching laboratory/hands-on skills at a distance. His review of the literature has determined that courses needing to teach to the psychomotor domain have some challenges. In physical sciences, engineering, and technical fields, laboratory studies have been considered key components of the curriculum (ABET, 2003). Current social and educational trends indicate the demand for the learner to be more flexible in participation times and settings (Moore, 2003). This may be accomplished through asynchronous activities. The explanations for distance education's current rising popularity also explain situations in the conventional classroom where learners increasingly justify absences with a growing list of unavoidable reasons. The flexibility of the distance education medium for activities such as posting class notes on a Web site for the truant resident student work well for the lecture, written examples, or testing portion of many courses; but what about the hands-on laboratory sessions that have been held in such high regard for many years? Holmberg and Bakshi (1982) suggest five alternatives to conventional laboratory arrangements:

- Eliminate laboratory attendance altogether.
- Make the laboratories available at times convenient to the student.
- Arrange for the use of local (relative to the learners' location) laboratories.
- Forward experimental kits directly to learners.

- Substitute laboratory work with alternative material (e.g., CAM and CAL packages, videos, etc.).

Laboratories are expensive, may be hazardous, require large areas per learner, have low learner-to-instructor ratios, receive low utilization if they are not adaptable to other courses, and compete inequitably for resources; but few in agriculture, engineering, science, and technology see an alternative. Many jobs mandate that learners have exposure, practice, and experience.

Agreeing that there are few alternatives to developing technical skills or acquiring familiarity with equipment, what about the cognitive understanding, the psychomotor skills, the application, analysis, and problem solving that are a part of laboratory goals? Are there alternatives for the conventional laboratory experience in these areas?

Members of the Accreditation Board for Engineering and Technology Education at their 2003 annual meeting considered distance education issues for technology and engineering, especially the challenges of laboratories (ABET, 2003). This accreditation group is still in the early stages of developing recommendations and standards for independent asynchronous studies utilizing distance education technologies in engineering and technical laboratories.

Trammell (n.d.) compiled a collection of bulletin board postings in response to questions about laboratories in distance education. In this collection, Kennepohl stated that they have no formal criteria for laboratory components, except that it should be equivalent to courses at traditional universities in Canada. Kennepohl goes on to state that the acid test is simply whether other universities will accept them as transfer courses. These seem to be common attitudes or assumptions.

Graham (1982) proposed a list of learning objectives for laboratory work with the three main issues being those of gathering knowledge relating to procedure, gathering knowledge of types of equipment, and analysis and reporting. Lemckert and Florance (2002) pointed out that these can only be fully achieved by learners participating in a broad range of interactive laboratory exercises conducted within a structured, suitably equipped environment. They go on to point out that the use of Internet mediated learning could satisfy all the aims of practical work except for the "hands-on" psychomotor skills; therefore, this type of training is likely best suited for the mature-age student who has significant experience with mechanical devices.

Writing Instructional Objectives

After considering the domain and level of instruction, it is time to write the instructional objective. We like to think of instructional objectives as a vehicle that helps deliver the learners from where they are to where they need (or want) to be. With that analogy in mind, consider that the vehicle has four wheels, representing four components (the ABCDs of objective writing): **a**udience, **b**ehavior, **c**ondition, and **d**egree. Who is this objective for? What will be measured? What are the conditions of the performance? What are the criteria or standards of performance? It may be like learning your ABCs, but it is as easy as 1, 2, 3! Follow these three steps to formulate your instructional objective statements:

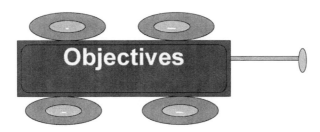

Step 1: Once the audience is defined, there is really no need to include it in the objective statement. That leaves the other three components. For learner-centered instruction, we like to use the term "observable action" rather than "behavior." Verbs are your actions, so refer back to the tables within the three domains and determine the verb that best describes what the learners will do. Each objective should only have one action verb and it should be specific, not vague.

	Poor	**Better**
Example	To know	To write
	To understand	To recite
	To appreciate	To compare
	To believe	To describe
	To enjoy	To construct
	To grasp	To solve

If there are several actions in the objective statement, it will be unclear and difficult to determine if the objective has been met. So break it down into one or more objectives (often called "enabling objectives") to support the primary instructional objective.

Step 2: Now think about the condition of the performance. What does the learner need in terms of instructional materials or activities to accomplish the observable action? This part of the statement identifies resources, procedures, materials, aids, tools, and so forth, to perform the task. The conditions can appear in any part of the objective statement and typically are expressed with a prepositional phase such as "after viewing a streaming video" or "without using the textbook."

Step 3: How well should the learner be able to perform this task given these conditions? This part of the instructional objective states the level of acceptable performance with quantity, quality, and time. Sometimes the assumption is made by the language used that the learner must perform perfectly. For example, think back to our example at the beginning of the chapter for geometry. *Given a scalene triangle* (condition), students will be able to *prove* (observable action) that the sum of the measures of the angles of a triangle is 180. Because this problem has one correct answer, then the learner either knows it or not. With adult audiences and more complex, higher-order thinking, there probably is not one "right" answer. How should competence of this type of material be measured? Through the use of authentic forms of assessment, the learner and instructor can determine the definition of "success" to enhance individualized lesson sequences or learner contracts. The use of grading or assessment rubrics for papers, projects, and portfolios are examples that would require such negotiation (to be discussed in chapter X). Defining "good" work from "poor" work requires assessment skills to ensure that materials are judged fairly in cases where grades are given or employees are required to demonstrate skill sets for job performance.

Here is an example. See if you can pick out the three parts of the instructional objective statement: *Readers of this textbook will be able to write an instructional objective using the three steps and observable actions from the text and tables provided.*

Audience   Readers of the textbook (see why it is not necessary to include—it is redundant)

Behavior   Step One: Observable action verb = Write

Condition   Step Two: Condition = Using the steps and observable actions in this chapter

Degree   Step Three: Tricky? Degree or criteria = An (one)

Remember the examples of the instructional objectives on geometry and welding? Now you try writing one:

Audience:

Domain of Learning: *Cognitive, Affective, Psychomotor* (choose one)

    Observable Action (Behavior):   Level:

    Condition(s):

    Degree/Criteria:

Complete Statement:

Refer to the application exercise for more practice on objective writing.

 Internet Connection
http://www.nwlink.com/~donclark/hrd/sat3.html

This reference is from the Big Dog Training Web site on Instructional System Design – Design Phase. A detailed description on writing objectives is provided as well as other valuable information about designing instruction.

Conclusion

A clearly stated instructional objective contains precise language that is measurable by the instructor and learner. Once the instructional objective is written, steps for designing instructional sequences and materials, and determining the appropriate delivery strategies will likely follow. If you consider the domains of learning and level within that domain and then follow the steps outlined in this chapter, you will be able to begin the first and most critical dimension of the instructional design process. Our next chapter will examine Gagné's Nine Events of Instruction and how to gain attention and stimulate motivation with your learners.

 Application Exercise

Choose an audience and topic area that you would like to convert or create for distance education delivery. Develop at least three instructional objectives for that audience/topic area. Consider the domains of learning: cognitive, affective, and psychomotor. At least one objective should require higher-order processes (at the upper part of the domain hierarchies).

References

Accreditation Board for Engineering and Technology (ABET). (2003). The challenge of instructional laboratories in distance education. Retrieved December 12, 2003, from *www.abet.org/AnnualMeeting/Presentations/Distance%20Ed-Rosa.pdf*

Bloom, B., & Krathwohl, D. (1956). *Taxonomy of educational objectives: The classification of educational goals, by a committee of college and university examiners. Handbook I: Cognitive Domain.* New York, Longmans, Green.

Brahier, D. (2000). *Teaching secondary and middle school mathematics.* Boston: Allyn & Bacon.

Graham, A.R. (1982). Obtaining maximum benefits from laboratory instruction. *Frontiers in Education Conference Proceedings, IEEE, 1982,* 148-151.

Harrow, A. (1972). *A taxonomy of the psychomotor domain. A guide for developing behavioral objectives.* New York: McKay.

Holmberg, R.G., & Bakshi, T.S. (1982). Laboratory work in distance education. *Distance Education, 3(2),* 198-206.

Krathwohl, D., Bloom, B., & Masia, B. (1964). *Taxonomy of educational objectives: The classification of educational goals. Handbook II: Affective domain.* New York: McKay.

Lemckert, C., & Florance, J. (2002). Real-time Internet mediated laboratory experiments for distance education students. *British Journal of Educational Technology, 33*(1), 99-102.

Mager, R.F. (1997). *Preparing instructional objectives: A critical tool in the development of effective instruction.* Atlanta, GA: Center for Effective Performance.

Moore, M.G. (2003). *From Chautauqua to the virtual university: A century of distance education in the United States.* Columbus, OH: Ohio State University, College of Education, Center on Education and Training for Employment. Retrieved December 18, 2003, from *http://ericacve.org/majorpubs2.asp?ID=37*

Newcomb, L., McCracken, J., Warmbrod, J., Whittington, M. (2004). *Methods of teaching agriculture.* (3rd ed.). Upper Saddle River, NJ: Pearson, Prentice Hall.

Trammell, G. (n.d.). Chemistry classes with at home labs. Retrieved November 1, 2001, from *www.uis.edu/~trammell/hom_lab.htm*

Chapter VIII

Events of Instruction:
Gaining Attention and Stimulating Motivation

 **Making Connections**

In previous chapters, we explored systematic instructional design, learner-centered instruction, and objective writing. Now we will give you some nuts and bolts on specific lesson planning and methods to gain attention and stimulate motivation in distance education. What are Gagné's Nine Events of Instruction and how do these events impact lesson planning? Why use icebreakers and openers in the lesson? How do you stimulate learner motivation? What kinds of things should be included in the closing segment of a lesson?

Introduction

You may recall in Chapters III and IV discussions about memory. Learners are constantly building mental models of the environment through experiences. The cognitive map provides a link between the thought process and the physical environment. About 95% of all new learning takes place through sight, hearing, and touch. Obviously, most of what comes in through the senses is sorted out very quickly through our perceptual or sensory registry. This process occurs in three to five seconds and must go into short-term memory for actual processing. Information that is transferred to short-term memory can remain active for about 15 to 20 seconds without rehearsal and generally has a limit of about five to seven items. We can think of short-term memory as a workbench area where we can build, take apart, or rework ideas for eventual storage. It is difficult to remember things for very long, such as a phone number we use for pizza delivery, unless we decide that the information is important.

- Does this information make sense?
- Can this information be understood based upon experience?
- Does the information fit into what is currently known about how the world works?
- Is it relevant?
- What is the purpose?

If the learner decides that the information presented makes sense and has meaning, then it is more likely to be stored in long-term memory (Atkinson & Shiffrin, 1968; Good & Brophy, 1986).

Thought and Reflection

MAKING A MEMORY

Memories are located throughout the brain. Explicit long-term memories are formed in the hippocampus while implicit long-term memories are formed elsewhere. Recall is found in specific locations in the cerebral cortex. The integration of facts and factoids blended with beliefs and experience, imbued with emotion, are combined and stored in different parts of the cortex forming the foundation of recall or reassembly as needed. Think of an important event in your life. Why can you remember details of this event more than less significant events?
Source: Zull (2002)

What does this brief introduction on memory have to do with lesson planning and gaining attention in distance education? Everything! Once the instructor moves from the planning stages of instructional design to the delivery of the program, the ultimate goal is to provide learning objects and authentic experiences to help the learner process and collect information (knowledge, skill, attitudes, or ability) for use at a later time.

 Internet Connection
http://chiron.valdosta.edu/whuitt/col/cogsys/infoproc.html

This site provides an overview of the information processing approach and an explanation on sensory memory, short-term memory, and long-term memory.

Learning Objects

High-quality instruction achieves its objectives with minimal cost and maximum effectiveness. One way to achieve this goal is through the use of learning objects. Specifically, learning objects are smaller chunks of instruction often stored in a database (rather than a full course or program) that can be used for a variety of settings and audiences. This helps to ensure integration, interoperability, and reusability. "The challenge lies in establishing useful standards and relatively seamless processes that can be readily adopted, implemented, maintained and improved by a critical mass of people and organizations" (Hirumi, 2003, p. 10).

Learning objects contain a measurable objective, an activity, and an assessment that are classified by metadata (Brennan, Funke, & Anderson, 2001). Metadata standards recognize that learning objects can have different attributes, such as document type, document format, didactical context, difficulty level, and interactivity level. While such attributes tell designers and instructors if they are dealing with an audio, video, text, or a graphic file, they do not require objects to contain fundamental instructional elements, such as objectives, activities, assessments, or feedback. Sequencing is difficult to create if you just string together a bunch of small learning objects. Objects assembled from various settings would have no pedagogical or andragogical relationship to each other. Learners who access objects directly from a database may find them lacking if they do not contain essential instructional elements (Hirumi, 2003). See Figure 1 for a diagram of the Sharable Content Object Reference Model (SCORM) to visualize the relationship of sharable content (learning objects) from the World Wide Web assembled through a server on demand and redistributed in various learning settings.

Learning objects (sharable content), such as an animation of protein synthesis or an annotated PowerPoint presentation on fungal spore infection of a leaf, will be more beneficial to instructors and learners if they are accompanied with details about their use and application.

Figure 1. ADL's long-term vision for use of sharable objects (Source: Advanced Distributed Learning, Sharable Content Object Reference Model (SCORM^TM) Version 1.2. Copyright 2001. Used with permission of the author.)

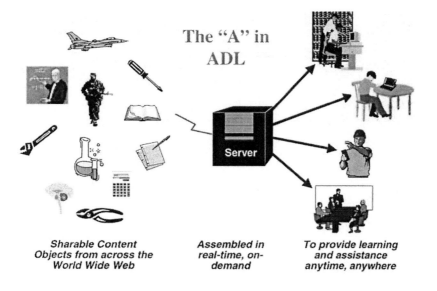

Gagné's Nine Events of Instruction

How do designers and instructors ensure that the appropriate instructional elements are present? We believe that it is necessary to create lesson plans to incorporate the objective, activity, and assessment components of the learning object, regardless of the media used (Morrison, Ross, & Kemp, 2004). Gagne's Nine Instructional Events is an effective way to plan lessons for distance education delivery.

You may recall the discussion on systematic instructional design in chapter VI. Gagné's work was influential on instructional design models. Merrill, Li, and Jones (1990) created a new design model entitled ID² or the Second-Generation Instructional Design Model based on Gagné's work. This model makes a strong link between learning outcomes and the internal/external conditions of learning. In the ID² model, performance is the result of cognitive structure or mental models. The construction of the mental model is facilitated by the instructional strategy and sequence and therefore promotes different learning outcomes (Merrill, Li, & Jones, 1990).

Smith and Ragan (1999) also elaborated on Gagné's theory, suggesting that the events of instruction do not consider learner-centered instruction. Smith and Ragan's model, called Comparison of Generative/Supplantive Strategy (COGSS), helps instructors, instructional designers, and learners determine the balance between instructional strategies and learning strategies based on context, learner, and task variable. It is evident that Gagné's work has impacted other instructional design models and theories of teaching and learning.

In Gagné's book, *The Conditions of Learning* (1965, 1970, 1977), he describes nine events of instruction that serve as guiding principles for designing instructional strategies. We believe that these events can still serve as guide-posts for lesson planning for distance education, if the instructor and/or instructional designer include opportunities for engagement and assessment centered on the learner. We will provide a synopsis and distance learning example of each event below.

- *Gain Attention (Reception)* – This event helps the learner prepare for the instructional events that follow. It consists of getting acquainted and setting the stage for instruction. The use of icebreakers/openers to help build rapport with the instructor and other learners is a great way to gain attention at the beginning of each learning object. Providing an overview of the equipment, software and/or plug-ins necessary to access the course content is another dimension of this event. The instructor can send a letter or e-mail to the learner before the course or training program begins to provide a logistical and technological orientation. Learners can also post a bio data sheet on the course Web page to help the instructor and other learners get to know each other and to bridge the transactional distance. A short survey about level of technology knowledge and skills possessed can also be given to serve as a needs analysis.

- *Inform the Learner of the Objective(s) (Expectancy)* – The second event involves making the learner aware of the instructional objectives. The instructional design process mentions the importance of developing objectives, so do not keep them a secret. This encourages the learner to focus on requirements, expectations, evaluation criteria, and materials to be covered. In a learner-centered environment, this can be negotiated with the learner to ensure applicability and transfer of training that is relevant. In a course delivered online, the objectives can be listed at the top of the Web page to help the learner gauge the focus of the content, necessary activities, and assessment components.

- *Stimulate Recall of Prior Learning (Retrieval)* – In a learning sequence, stimulating recall allows the learner to scaffold or connect old knowledge or skills to help store new information in long-term memory. For example, if teaching or training on a topic with multiple lessons or themes, ask the learners questions pertaining to the previous lesson and tie them into the new content. This can be done as text on the Web page after citing the objectives, in a discussion forum, or through an opening video or audio segment to introduce the new material.

- *Present the Content/Provide Stimuli (Selective Perception)* – For this event, it is important to emphasize active techniques, ones that will help the learner retain the new knowledge, skills, or abilities (competencies), and encourage critical and creative thinking, interactions, and problem solving. It is important to chunk or group content in a meaningful way, and to use a variety of media types to accommodate different learning preferences. A learning object should stand alone. The content should be independent of the media and portable to allow application in other contexts and through other media formats.

- *Provide Learning Guidance (Semantic Encoding)* – This is the event that helps learners store the information in long-term memory so that it can be retrieved later. Activities such as guided note taking, questions, discussion, follow-up activities, interactive study guides, and so forth, provide opportunities for learning guidance. Provide advanced organizers, such as printable copies of your presentation materials and instructions for small group or virtual team (collaborative learning) assignments.

- *Elicit Performance (Response)* – This event allows the instructor/facilitator to determine if the learner has acquired the necessary competence based on the instructional objectives. If the objectives were negotiated, as suggest by Hirumi's SCenTRLE model (Student-Centered, Technology-Rich Learning Environment, 2002) previously discussed in chapter VI, then meaningfulness and relevancy will be enhanced. This also promotes self-directedness and stimulates motivation for the learner.

- *Provide Feedback (Enforcement)* – Gagné contends that this event helps the instructor/facilitator determine if the intended objective (learning outcome) can be consistently performed and reinforces competencies. Based on SCenTRLE, we believe that providing feedback is an integral step throughout the learning process and should be a component of every learning activity. This requires special facilitation skills (and time management), which will be discussed more fully in the next chapter.

- *Assess Performance (Retrieval)* – This event is an extension of the previous event in that it determines learner performance based on outcomes. With SCenTRLE, the learner would develop an evaluation rubric and have expert authenticators determine if the process/product reached the desired learning objectives. On a lesson planning level, daily or weekly performance checks may take on a more informal role to include an e-mail confirmation or electronic discussion of progress. We will discuss assessment strategies in chapter X.

- *Enhance or Reinforce Retention and Transfer (Generalization)* – And finally, the ninth event provides cues/strategies and practice of the newly learned knowledge, skills, or abilities to ensure that they are retained and can be applied to new situations. This is extremely important when working with adult audiences. By negotiating learning objectives and outcomes, chances improve that the learner will create an experience that will transfer to his or her work or educational setting. This event also means that review and testing may take place, summaries of the lesson are developed, conclusions are drawn, plans for application are developed, and next steps are outlined as appropriate.

Internet Connection

http://www.e-learningguru.com/articles/art3_3.htm

This site provides methods to apply Gagné's Events of Instruction in distance education settings.

Icebreakers and Openers: Why Use Them?

The first event of instruction is *gaining attention*. This can be done through the use of icebreakers and openers. Think of it as an IOU—**I**cebreakers and **O**peners increase **U**nderstanding. People tend to remember things in threes and most people can remember acronyms. You might think the IOU will be money or a nice prize and it will probably pique your interest (or extrinsic motivation). This is one example of a simple technique to focus attention prior to the delivery of new content.

According to *The Winning Trainer* (Eitington, 1996), there are two ways to start a course or training program: 1) introduce the group to the content or 2) ease the group into things before directly involving them in the content. An icebreaker is used to ease participants into the learning experience before involving them directly in the content. Icebreakers are typically used to reduce fears, tensions, and anxieties; energize the group; and set the tone for the program. In contrast, an opener is used to introduce participants to the content at the outset of the experience. You only have one chance to make a good first impression. Whatever you do sets the stage for your philosophy, style, confidence, and competence. See Figure 2 for examples of online icebreakers.

Why should instructors use icebreakers and openers? Icebreakers and openers can serve as an audience analysis and a means to build rapport between and among the instructor and learners. For example, learners can provide a short biographical sketch about themselves and what they would like to learn during the course or training program. This helps the learners get to know each other, helps the instructor determine learning goals and interests, allows the learners to use the communication technologies in a nonthreatening way, and establishes a climate for interaction. This is true in any instructional setting, but even more

Figure 2. Online icebreaker ideas

Online Icebreaker Ideas

Two Truths and a Lie – Have participants send two truths and a lie over a course listserv or threaded discussion list (either whole class or in smaller virtual teams). Have the other learners respond with their guess of the lie.

Costume Party – Have each student choose a costume to represent a personality trait. They can take a digital picture and send it to the other learners with a description.

Write Your Own Epitaph – Participants will think about where they think life will lead them (goals and aspirations) and then write their epitaph as others would view their contributions.

Pass It On – Each participant will develop a question to pass on to the next participant (e.g., "Where was the tallest tree you have ever climbed?"). The participant then e-mails the question to another participant. The second participant answers the question and then e-mails the response along with the original question on to the next person. The process is repeated until everyone has had a chance to answer. Then the responses are returned to the originator of the question.

Name Game – Have each learner create a mnemonic of their first name and use that to describe themselves. Here's an example. My name is Corky and I have a crazy sense of humor and love to have fun. My hobby is amateur radio. My wife is much younger than I and we have a 4-year-old. I teach a freshman "Orientation to College" course, and as an "old" guy (compared to my students) I enjoy the company of youngsters.

[C] Crazy- [O] Old-[R] Radio ham- [K] Knows- [Y] Youngsters

so in distance education. During the opening segment of the lesson, you will want to warm up or energize the participants, inform the learners of the objective(s), stimulate recall of previous lessons (if applicable), and provide advanced organizers to help the learners visualize the path of instruction. Even if you are not confident thinking up your own icebreakers or openers, a variety of Web sites and resources are available to assist you. Most are designed for face-to-face settings, so you may need to adapt them for distance learning.

Using icebreakers and openers helps to build knowledge and make links to past information. It is also an opportunity to vary the pace and learning style preferences. When planning the opening for the lesson, you should consider the

Internet Connection

http://www.resultsthroughtraining.com/downloads/Icebreakers.html

This site links to examples of icebreakers that can be used in training settings. Think of ways to modify a few of them for distance education settings.

composition and expectations of the group, nature and length of the program, culture of the sponsoring organization, and style and personality of the instructor (Eitington, 1996). When making your selection of the icebreaker or opener you will use, think about the amount of time it will take, the novelty of the approach, the creative quality, and whether it will be interesting and exciting, or threatening to the participants.

Stimulating Motivation

Based on Gagné's Nine Events of Instruction and the use of icebreakers and openers, let us now consider instructional strategies for stimulating motivation and interaction. Motivation theory is not considered to be its own theory of instruction. Motivation theory does fit within the models of instruction that deal with conditions, strategies, and goals.

There are four major dimensions of motivation within instructional theory: (1) interest, (2) relevance, (3) expectancy, and (4) satisfaction. Interest can be described as stimulating the learner's curiosity. Relevance is determined by whether the instruction satisfies the learner's personal needs or goals. Expectancy determines the learner's perceived likelihood of success and feeling of learner control. Satisfaction is determined by how well the learner enjoyed the learning experience.

There is a belief that motivation results from the interaction of various reinforcers that are positive (carrots) and/or negative (sticks). Grades are often the sticks and carrots of the formal classroom. Grades reflect motivational qualities, such as self-discipline and competitiveness, in addition to academic achievement. When working with adults, what are the differences between intrinsic (self-defined) and extrinsic (externally defined) motivation? External forces are tangible rewards while intrinsic forces include personal satisfaction, feelings of self-determination, and competence.

Instructors most skillful at motivating learners recognize the great variations in academic abilities, interests, and attitudes. Because extrinsic motivators are powerful and widespread, learners are influenced by rewards such as grades. Instructors who deemphasize grades and encourage intrinsic motivations must consider activities and evaluation strategies to stimulate learning. Learning does not occur as a result of the design and delivery of the media, but rather by what the learner *does* with that media. True meaning and understanding of the instructional content takes place when the learners are immersed in the content through engagement and active participation.

Driscoll (1994) noted four components of motivation and corresponding strategies for each. The first is *gaining and sustaining attention*. We discussed capturing the learners' attention in the previous sections of this chapter. Other approaches include stimulating lasting curiosity with problems that invoke mystery and maintaining attention by varying the instructional presentation and by presenting problems being faced by the learners that they want to solve. The second component of motivation is *enhancing relevance*. Strategies to enhance relevance include increasing the utility of instruction by stating (or having the learners determine) how instruction relates to personal and professional goals; providing opportunities for learners to match their motives and values with occasions for self-study, leadership, and cooperation; and increasing familiarity of instruction by building on learners' prior knowledge and abilities. The third component of motivation is *building confidence*. Strategies that create a positive expectation for success by making instructional goals and objectives clear and provide learners opportunities to attain goals successfully and have a degree of control over their own learning are most effective. The fourth component is *generating satisfaction*. Create opportunities for learners to use newly acquired skills and use positive reinforcement to complement their skill development. Also, ensure that standards of assessment are consistent and match with the outcomes designated in the course objectives.

These components are important reminders to put into practice. We will explore various media that can facilitate this process in Section IV of the book. As you begin to design lessons for delivery at a distance, consider that learners who are motivated can learn regardless of the delivery strategy or medium used (Heinich, Molenda, Russell, & Smaldino, 1996). A summary is provided in Table 1.

Table 1. Instructional strategies for stimulating motivation (Source: Driscoll, 1994, p. 319)

Component of Motivation	Corresponding Strategies
Gaining and Sustaining ATTENTION	• Capture students' attention by using novel or unexpected approaches to instruction • Stimulate lasting curiosity with problems that invoke mystery • Maintain students' attention by varying the instructional presentation
Enhancing RELEVANCE	• Increase the perception of utility by stating (or having learners determine) how instruction relates to personal goals • Provide opportunities for matching learners' motives and values with occasions for self-study, leadership, and cooperation • Increase familiarity by building on learners' prior knowledge
Building CONFIDENCE	• Create a positive expectation for success by making clear instructional goals and objectives • Provide opportunities for students to successfully attain challenging goals • Provide learners with a reasonable degree of control over their own learning
Generating SATISFACTION	• Create natural consequences by providing learners with opportunities to use newly acquired skills • In the absence of natural consequences, use positive consequences such as verbal praise, real or symbolic rewards • Ensure equity by maintaining consistent standards and matching outcomes to expectations

How to End the Lesson

Just as important as gaining attention in the beginning, the lesson sequence should have a definite closing and component that assures that the learning can be applied to other settings. Learners need meaningful opportunities to apply their learning to work or life in general. The activities used throughout the lesson or course should culminate in an exercise that allows the learners to synthesize the content and make application. Activities such as observation assessment, practical assignments, role plays, postsession action plans, and contracts can help ensure skill retention and transfer (Eitington, 1996). The closing should also include a wrap-up and review of content delivered and opportunities for evaluating the course and its delivery strategies. It is important also to draw conclusions about the content of the lesson. For example, in teaching a unit on characteristics and effects of different types of fertilizers, the students should be able to conclude that sodium nitrate fertilizers should not be applied to alkaline, clay soils. Formative and summative evaluation techniques and authentic assessments will be covered in chapter IX.

Conclusion

Now that we have examined the events of instruction, how to open and close our lesson, and how to stimulate motivation in our learners, we should have the necessary elements to put these strategies into practice. One suggestion is using a lesson planning template. An excellent template can be found at Results Through Training provided in the Internet Connection section below. Because you have developed objectives in the previous chapter, take a few moments to complete the application exercise.

 Application Exercise

As you start to plan a lesson for delivery, we suggest the Results Through Training template as a guide.

http://www.rttworks.com/downloads/desgndoc.rtf

Download this form on your computer and replace the introduction section with your own materials. In the first chunk called "Introduction," select or create an icebreaker or opener that you would use in a distance education course or training program to gain attention and provide an overview to your topic. Describe the method/activity, content, support materials, and estimated time. Also include your objectives developed from the previous chapter.

References

Atkinson, R., & Shiffrin, R. (1968). Human memory: A proposed system and its control processes. In K. Spence & J. Spence (Eds.), *Psychology of learning and motivation: Advances in research and theory* (Vol. 2, pp.89-195). New York: Academic Press.

Brennan, M., Funke, S., & Anderson, C. (2001). The learning content management system: A new e-learning market segment emerges. Retrieved November 9, 2001, from *www.avaltus.com/idc/index.html*

Driscoll, M.P. (1994). *Psychology of learning for instruction.* Upper Saddle River, NJ: Pearson/Allyn & Bacon.

Eitington, J.E. (1996). *The winning trainer: Winning ways to involve people in learning.* Houston, TX: Gulf.

Gagné, R.M. (1965). *The conditions of learning* (1st ed.). New York: Holt, Rinehart and Winston.

Gagné, R.M. (1970). *The conditions of learning* (2nd ed.). New York: Holt, Rinehart and Winston.

Gagné, R.M. (1977). *The conditions of learning* (3rd ed.). New York: Holt, Rinehart and Winston.

Good, T.E., & Brophy, J.E. (1986). *Education psychology: A realistic approach.* New York: Longman.

Heinich, R., Molenda, M., Russell, J., & Smaldino, S. (1996). *Inst. media and technologies for learning.* Up. Saddle Rv, NJ: Pearson/Prentice-Hall.

Hirumi, A. (2002). Student-centered, technology-rich, learning environments (SCenTRLE): Operationalizing constructivist approaches to teaching and learning. *Journal for Technology and Teacher Education, 10*(4), 497-537.

Hirumi, A. (2003). *A new system for e-learning. A draft white paper.* Unpublished manuscript.

Merrill, M.D., Li, Z., & Jones, M.K. (1990). Limitations of first generation instructional design. *Educational Technology, 30*(1), 7-11.

Morrison, G.R., Ross, S.M., & Kemp, J.E. (2004). *Designing effective instruction* (4th ed.). Hoboken, NJ: John Wiley & Sons.

Smith, P.L., & Ragan, T.J. (1999). *Instructional design* (2nd ed.). New York: John Wiley & Sons.

Zull, J.E. (2002). *The art of changing the brain: Enriching the practice of teaching by exploring the biology of learning.* Sterling, VA: Stylus.

Chapter IX

Learner-Centered Assessment and Facilitation Techniques

with
Barry Boyd, Texas A&M University, USA,
Kathleen Kelsey, Oklahoma State University, USA, and
Atsusi Hirumi, University of Central Florida, USA

 Making Connections

You have just finished grading the first exam and the results are in. What does the grade really mean? Did the participants of the course or program learn? How do you know? How will you assess learning at a distance? A fundamental step in systematic instructional design and delivery is deciding how to assess learning outcomes. A well-written instructional objective includes outcome measures. Making certain that the assessment strategies match the objectives is an important first step, a step that must not be overlooked. A frequent concern raised by distance educators is that online delivery takes more time to facilitate and assess than traditional classrooms. In this chapter, we will discuss the use of formative evaluation and authentic assessment techniques to determine instructional effectiveness and learning outcomes. To establish viable online programs, we need to optimize the amount of time educators spend online. Tactics for optimizing time spent facilitating online learning will conclude the chapter.

Introduction

Educators are always being asked two basic questions. Do you know what they are? You don't have to read a report by the National Commission of Excellence in Education (1984) to find out. What do employers, parents, and personnel in state accreditation agencies want to find out from you? Aren't they asking (1) How well are learners doing? and (2) How effectively are you (as instructors) teaching? How should we measure what and how much our learners understand? There is public and political pressure to explain learning. Through observation of the process of learning, the collection of feedback on learning, and the design of modest experiments, instructors can determine how learners respond to particular instructional approaches (Angelo & Cross, 1993).

Assessment refers to the broad area of monitoring the learning progress. Assessment is the umbrella term that covers a variety of data collection methods used to evaluate educational outcomes (Chase, 1999). It should occur both during instruction (formative) and at the end of instruction (summative). Angelo and Cross note four characteristics that define assessment (1993). First, assessment is learner-centered; its purpose is to assure student learning. Formative assessment is used to determine if learning is occurring during the process of instruction. It involves both the instructor and the learner and is mutually beneficial to each. If the instructor discovers that learners are struggling with the material, adjustments may need to be made with either the instructional methods or the processes of how the student goes about learning. Assessment can provide the information necessary to guide the instructor and learner in making those adjustments.

Assessment is instructor directed. It depends on the knowledge and professional experience of the instructor to determine who and what needs to be assessed. Assessment must be tailored to meet the needs of the learners, the instructor, and the content being taught. It is important to choose the right assessment technique to fit the situation.

Assessment must also be an on-going process. Multiple, simple assessment techniques will provide the instructor with sufficient information to alter instructional methods if necessary. Feedback can also be provided to the learners on how they can improve learning.

Finally, assessment builds on existing best practices by making it more systematic and more effective. Integrating assessment into the lesson plan makes it a seamless part of the course program.

Formative Assessment Techniques

Many of us are accustomed to traditional assessment methods which do not provide the kind of continuous feedback necessary for continuous improvement (Huba & Freed, 2000). Through exams and grades we provide messages to our learners but it only serves to *monitor* learning, not *promote* it. The purpose of formative assessment is to help instructors direct and redirect their instruction to improve learning. There are often gaps between what was taught and what has been learned. By the time an instructor notices these gaps in knowledge and understanding, it may be too late to fix the problems. Formative assessment may point to a need for adjustment to the instructional strategy, and it can help identify learners who have not grasped the idea or concept.

Formative assessment techniques allow the instructor to make immediate changes in the course or program. To determine which assessment technique is appropriate to use, the instructor should follow three steps: 1) decide which technique will provide the information needed, 2) implement the technique, 3) respond to the feedback collected by making changes, if indicated, and providing feedback to the learners (Huba & Freed, 2000). Following are several assessment techniques that can provide formative assessment for the instructor and learners (Boyd, 2001).

The Minute Paper

Angelo and Cross (1993) describe the minute paper as one of the most frequently used assessment techniques. It is simple to use and provides immediate feedback on the learner's grasp of the content. At the end of a videoconference or instructional sequence online, ask the learners to respond to the following two questions:

1.   What was the most important thing you learned during this session?
2.   What important question remains unanswered? (Angelo & Cross, 1993, p. 148).

Learners have one minute to write their responses. Questions can be changed to collect any type of information the instructor requires, but the goal remains to have students respond to a few questions in a short period of time.

Muddiest Point

This purpose of this assessment tool is to identify and clear up any confusion about the lesson content before new material is introduced. This tool only involves one question: What was the muddiest point in the lesson? The instructor may pose this question about an assignment, a project, a demonstration, or any other learning object. If several students describe the same points, the instructor may need to modify how the information is presented to improve learning (Huba & Freed, 2000).

One-Sentence Summary

The purpose of the One-Sentence Summary is to determine how well learners can summarize information about a particular topic. The instructor focuses on a particular concept and asks the students to describe who does what to whom, when, where, how and why (WDWWWWHW) (Huba & Freed, 2000). Learners should answer the questions in one long sentence. This technique encourages learners to focus on key questions when they read an assignment or participate in an activity.

Application Cards

Students enjoy learning if they can see how concepts can be applied to real-world situations. The application card activity helps students learn multiple ways that lesson content can be applied. Following a lesson, learners are asked to write down on an index card at least one real-world application for what they have just learned. This helps the learners connect new knowledge to its application (Huba and Freed, 2000). These can be used as a reflection exercise for the learners or shared directly with the instructor or other learners through email or threaded discussion online.

Approximate Analogies

Approximate Analogies allow an instructor to determine if learners understand the relationship between two ideas or concepts. Learners simply complete the analogy – A is to B as X is to Y. The instructor supplies the first half of the analogy and the learners supply the second half. A distance education example might be, "design is to teaching as _____ is to _____." An example from a communications course might be, "body posture is to communication as _____ is to _____." The analogies can be sorted into "good, questionable or wrong." Instructors choose samples to send back to the class over email or class list, explaining why each sample fell into its respective category (Angelo & Cross, 1993, p. 193).

Turn to Your Partner

Instructors can acquire feedback regarding learners' understanding of the lesson content by using a technique called Turn to Your Partner (TTYP). Students gain a deeper understanding of the material by first reflecting on the material and then exchanging their thoughts with other students. The instructor first poses a question about the material, instructing the learners to formulate their own answer first. The instructor then asks them to "turn to your partner." The pair discusses each other's answer to the question and formulates a new answer together. The instructor then calls on pairs at random to share their answer, thus gaining valuable feedback about the collective understanding of the material (Huba & Freed, 2000). This works very well in a videoconference by muting the system for a specified period of time. It can be modified for online instruction by using virtual teams or small groups through discussion forums.

Authentic Assessment

The use of formative evaluation strategies helps us to gauge how well *instructional strategies* are working (Dooley, Richards, & Lindner, 2002). There is also a need to develop and refine *assessment strategies* to evaluate and authenticate learning. Some courses may use traditional assessment measures, and Web-based software typically includes online quizzing/testing features. Quizzing software allows the instructor to create test banks from which to select questions randomly for the learner, grade them, and insert the score in the grade book. These items most frequently assess low-level knowledge and are *indirect* indicators of more complex abilities (Huba & Freed, 2000). Tests typically have one *right* answer. However, the challenges faced by adults "tend to be those that require the simultaneous coordination and integration of many aspects of knowledge and skill in situations with few right answers" (Huba & Freed, 2000, p. 12). As we shift to learner-centered instruction, "we should design assessments to evaluate learners' ability to think critically and use their knowledge to address enduring and emerging issues and problems in their disciplines" (Huba & Freed, 2000, p. 12).

Most people think of standardized, formal testing when assessment comes to mind; however, several forms of authentic assessments (papers, projects, and portfolios) are excellent ways of documenting the learning process. We suggest those three Ps: papers, projects, and portfolios. Why? This type of assessment is particularly powerful for instructors and learners at a distance (Dooley, Edmundson, & Hobaugh, 1997).

Distance education has typically taken advantage of advances in communications technologies. The primary medium available for learner interactions has been the written word (Murphy, Lindner, & Kelsey, 2002). Information technologies have made it possible to provide learners with additional media channels. The telephone, audio tape recorders, and video cameras have been widely available for decades. Desktop videoconferencing is another common technology. These technologies have not replaced the written word as the preferred medium for learner interactions. The latest technologies, from instant messaging (IM) to blogging, rely on written expression. Murphy, Lindner, and Kelsey (2002) argue that the written word will be the dominant medium for learner interactions in the near future. Writing competence is fundamental for success in distance education environments (Murphy, Lindner, & Kelsey, 2002).

Papers

Conducting authentic assessments of writing can be problematic. Writing samples include reporting a news event, narrating a story, critiquing an argument, and revising a memo to name a few (Breland, Bridgeman, & Fowles, 1999). Assessing learners' writing competencies is complex because of the diverse types of writing samples. Subjective assessment can cause frustration for instructors and learners. Assessments of writing should be "appropriate for a particular purpose and population" (Breland, Bridgeman, & Fowles, 1999, p. 1).

One suggestion for evaluating writing is to use assessment rubrics. A rubric is a criteria rating scale, which gives the instructor a tool to track learner performance and gives the learners a way to judge their work (Hirumi, 2003; Huba & Freed, 2000). For example, a sentence structure assessment rubric might include six writing competencies (coherence, audience awareness, argument, summary, sources, and grammar) (Murphy, Lindner, & Kelsey, 2002). In a study conducted by Murphy, Lindner and Kelsey (2002), coherence is the development of a clear thesis and introduction and the inclusion of well-constructed paragraphs. Audience awareness is the ability to write on the appropriate level for the intended audience using correct tone and voice. Argument is the development of a logical argument supported with important consequences. Summary is the development of a clear conclusion restating the main premises and drawing on the logical argument and references. Sources are the appropriate use of references following a citation format such as that of the American Psychological Association (APA). Grammar is the ability to write a paper free of grammatical errors. In the example rubric below (Figure 1), there are four possible scores on overall writing strength: 4 = demonstrates adequacy; 3 = suggests adequacy; 2 = suggests inadequacy; and 1 = demonstrates inadequacy. Also, the Internet Connection provided gives information about designing your own rubrics.

 Internet Connection
http://jonathan.mueller.faculty.noctrl.edu/toolbox/rubrics.htm

This site provides a useful reference to help you design your own rubrics for authentic assessment.

Projects

Another type of authentic assessment is to use project-based learning. A project can be defined as an assignment over a prolonged period of time that requires setting goals, planning, using resources, organizing, making judgments, and crafting a written and/or visual presentation of material or other outcome. Projects can be completed alone or by working with others (Huba & Freed, 2000). According to the Project-Based Learning Guide, project-based learning is learner-centered and emphasizes learning activities that are long term,

Figure 1. Example of a rubric on assessment of writing drawn from Murphy, Lindner, & Kelsey (2002)

Competency Category	*Demonstrates Adequacy (4)*	*Suggests Adequacy (3)*	*Suggests Inadequacy (2)*	*Demonstrates Inadequacy (1)*
Argument – Development of a supported and logical argument about an issue with important consequences for both the author and the audience				
Coherence – Development of a clear thesis and introduction that sets the stage for the argument and well-constructed paragraphs in the body of the text				
Grammar – Ability to write a grammatically correct paper				
Summary – Development of a clear summary that draws on the established argument and references				
Audience Awareness – Ability to write paper on appropriate level for identified audience and to make appropriate appeals using correct tone and voice				
Sources – Uses appropriate references in the paper following APA documentation				

interdisciplinary, and integrated in real-world contexts. Project-based learning allows learners to pursue their own interests and use critical thinking and problem-solving strategies to complete the task. It also promotes the application of multiple intelligences and helps to build motivation and a sense of accomplishment, giving the learner a sense of ownership and control over his or her own learning.

Internet Connection
http://www.4teachers.org/projectbased

This site provides links to checklists, rubrics, and reasons for using project-based learning.

Portfolios

Some instructors use a portfolio to catalog project-based learning or other holistic approaches. A portfolio is a purposeful collection of work that exhibits the learners' process, progress, and achievements. The portfolio should include the criteria for judging merit and evidence of self-reflection (Paulson, Paulson, & Meyer, 1991). The real power of portfolio assessment is that it gives learners an opportunity to *learn about learning* (metacognition) through engaged self-reflection. A portfolio should be created *by* the learner. Instructors can provide criteria for assessment, offering learners a concrete way to value their own work. A portfolio should include a rationale, goals, contents, standards, and judgments. This organized and hopefully relevant material will be particularly important to the instructor for providing feedback to the learners, to show growth, and to reveal problem areas that need to be addressed. Learners should be given examples of portfolios to serve as a guide.

 Internet Connections
http://www.ash.udel.edu/ash/teacher/portfolio.html

The site above provides definitions and examples of electronic portfolios. If you would like to read an article on the topic, check out the link below from *T.H.E. Journal* entitled, "A New Wave in Assessment" by Anna Maria D. Lankers.

http://www.thejournal.com/magazine/vault/A3380.cfm

Why use the three Ps (papers, projects, portfolios) in conjunction with, or in lieu of, testing? Standardized testing may distort the educational process because instruction is often driven by the testing process. The three Ps can help learners understand the link between what they study and their future successes by integrating learning and assessment. Therefore, the three Ps can complement traditional testing and provide educators with new tools to measure learning. The three Ps demonstrate progress over time in a more natural way and aid in the development of instructional content to guide course or program revisions. Learners also benefit by fostering personal responsibilities, improving self-esteem, developing critical-thinking skills, and gaining a sense of pride and ownership of their work.

The three Ps use competencies to document learning as measured by a set of criteria, such as behavioral benchmarks. Often these criteria are associated with a number so that instructors can manage the evaluation process. This type of assessment allows an instructor to maintain information about a learner's performance without losing the qualitative information indicative of other types of performance and growth over time.

In distance education, we can provide an environment where learners may be evaluated through these types of authentic assessment. For example, the portfolio can be project-based and available in electronic format, providing not only a synthesis of the course content, but also integration of technology skills. Learners can create the portfolio as a Web page, presentation, videotape, or other media type that can also be used to demonstrate skills when applying for professional positions or to document professional development.

Time-Saving Techniques for Online Instruction

Courseware makes it easier to facilitate and assess online instruction. However, courseware alone does not transform traditional teaching materials into effective online courses nor does it prevent educators from being inundated by e-mail and bulletin board postings. Ease of use does not necessarily translate into the development of innovative environments that use the potential of emerging technologies to stimulate collaborative and individual learning. The problem lies in applying instructor-directed instructional methods that are inappropriate for facilitating active online learning (Hirumi, 2003).

Hirumi (2003) provides five optimizing tactics to help instructors manage their time in facilitating online instruction. These tactics also have implications for providing feedback during formative evaluation and authentic assessment.

Align and Publish Objectives and Assessments

Learning objectives and assessments are common elements of the instructional process. Few, however, seem to approach these two vital elements in a systematic fashion. The alignment of objectives and assessment criteria is fundamental to high-quality instruction (Berge, 2003; Dick, Carey, & Carey,

2001). Learners should be informed of what they are expected to know and be able to do and they should be evaluated using assessment items and published performance criteria that match the behaviors specified in the objectives. If an objective states that learners will be able to *list* key concepts, assessments should ask learners to *list* key concepts. If an objective states that learners will be able to *compare* cases, assessment should ask learners to *compare* cases. It cannot be emphasized too much that this alignment between instructional objectives and assessments of learning must occur if authentic assessment is to take place and if the resulting assessment is to be credible, realistic, useful, and informative.

Align Instructional Events to Support Objectives and Assessments

To facilitate online learning, assignments and activities (instructional events) should also be aligned to support the achievement of specified objectives (Berge, 2003; Smith & Ragan, 1999). Learners need opportunities to practice and develop skills and abilities specified in the objectives. Classify objectives to a particular learning domain (cognitive, affective, or behavioral/psychomotor) and ensure that instructional events promote the achievement of the objectives within that domain (refer to Chapter VII).

Analyze and Balance Interactions

As was mentioned in the previous chapter, too many interactions may impede, rather than facilitate, learning. In online instruction, the instructor must acknowledge receipt of assignments, save and store files, print out, review, and assess each document, generate and distribute detailed feedback, and make sure learners understand the feedback. Considering the time and effort necessary to complete each interaction, it is easy to see why distance educators feel overwhelmed during the delivery of an online course. Therefore, it is important to examine the frequency and nature of planned learner–human and learner–nonhuman interactions and only include those that contribute to the learning objectives. Consider the use of virtual teamwork and group assignments to help manage the amount of feedback and provide opportunities for learners to engage with one another.

Create Feedback Templates

Feedback is vital to learning. It can be used to increase response rates or accuracy, reinforce correct responses to prior stimuli, or change erroneous responses (Kulhavy & Wager, 1993). Telecommunications expand our feedback options: e-mail, bulletin boards, chat, audio and video conferencing, and so forth. Without feedback, instruction may become "passing on content as if it were dogmatic truth, and the cycle of knowledge acquisition, critical evaluation and knowledge validation, that is important for the development of higher-order thinking skills, is nonexistent" (Shale & Garrison, 1990, p. 29). Without feedback, one-way ineffective and often inaccurate communication is taking place with the result that the instructor simply does not know why the learners performed as well or as poorly as they did. From a learner's perspective, it is often the ability and commitment to provide immediate, detailed, and appropriate feedback that distinguishes a good online educator. An effective feedback template consists of four primary components: (1) assessment criteria, (2) confirmatory feedback, (3) corrective feedback, and (4) personalized messages.

Assessment criteria are published to help guide learners' efforts, to assess their work, and to provide both confirmatory and corrective feedback. Confirmatory feedback lets learners know what they did correctly. Corrective feedback identifies areas that need improvement, provides insights on how learners can revise their work, and provides reasons for the suggested revisions. For example, scan all the assignments. Select what appears to be the most inadequate work sample and grade it first. Prepare a list of corrective comments that includes the rationale for the comments. Then select what appears to be a model assignment and grade it to generate a list of positive confirmatory comments. Use both lists to create an initial feedback template that consists of the published assessment criteria and a list of confirmatory and corrective comments. This allows you to copy and paste appropriate comments rather than having to generate them each time. Personalized messages integrated with the confirmatory and corrective feedback statements can be added to refer to unique aspects of the learners' work and demonstrate that you took the time to read and review properly each assignment.

Establish Telecommunication Protocols

The use of interactive technologies does not guarantee that meaningful interactions will occur. Establish protocols with guidelines and performance requirements for posting on threaded discussions, chats, virtual teams, and so forth. To optimize the value of threaded discussions, learners must actively participate in and take responsibility for the discussions. They must have a clear purpose and the skills necessary to use the system. The discussions must be well organized and easy to follow. For example, post a description of the activity with the purpose and rationale for participating. This description should include guidelines and performance criteria, a series of questions used to initiate the discussions, and an example of what is considered best practices (see Internet Connection below). Once an instructor has established guidelines, allow the learners to serve as facilitator and guide interactions for other topics. The role of the instructor, then, is to review, assess, and respond as necessary.

 Internet Connection

http://www.westga.edu/~distance/ojdla/spring51/edelstein51.html

This site provides a link to an article written by Susan Edelstein and Jason Edwards entitled, "If You Build It, They Will Come: Building Learning Communities Through Threaded Discussion." It is an excellent reference on evaluating learner participation in online discussion.

Conclusion

Instructors often teach the same way they were taught, usually through the use of lecture, demonstrations, and labs, to disseminate information and then give tests to assess learning (McMahon, 2000). As the field of education shifts from an instructor-centered philosophy to accepting and practicing a more learner-centered philosophy, the use of formative evaluation and authentic assessment becomes increasingly important. The assessment techniques in this chapter can help instructors determine if their instructional methods are effective and what, if anything, can be changed to improve learning. All of this takes time, so strategies to optimize facilitation techniques will help instructors manage online feedback, assessment, and communication to create the most efficient and effective learning communities.

 Application Exercise

Instructors and designers need to consider specific and measurable learner outcomes (assessment) to determine if their objectives have been accomplished. Design and develop the assessment instruments you will use to determine if your lesson, and the online delivery of your lesson, accomplished what you intended to accomplish (your objectives). Your assessments must match each of your instructional objectives. For example, you may determine that some of your objectives might best be measured by giving your audience an online test or quiz. Or some of your objectives might best be measured by having your audience produce a project or presentation, write a paper, or design an electronic portfolio. Be creative. Explore a variety of possibilities. Even create your own rubric!

References

Angelo, T.A., & Cross, K.P. (1993). *Classroom assessment techniques.* San Francisco: Jossey-Bass.

Berge, Z.L. (2003). Active, interactive and reflective e-learning. *Quarterly Review of Distance Education, 3*(2), 181-190.

Boyd, B.L. (2001). Formative classroom assessment: Learner focused. *The Agricultural Education Magazine, 73*(5), 18-19.

Breland, H.M., Bridgeman, B., & Fowles, M. (1999). *Writing assessment in admissions to higher education: Review and framework* (ETS Research Report No. 99-3). Princeton, NJ: Educational Testing Service.

Chase, C.I. (1999). *Contemporary assessment for educators.* New York: Addison-Wesley.

Dick, W., Carey, L., & Carey, J.O. (2001). *The systematic design of instruction* (5th ed.). New York: Addison-Wesley.

Dooley, K.E., Edmundson, C., & Hobaugh, C. (1997). Instructional design: A critical ingredient in the distance education soup. In L.M. Dooley (Ed.), *Distance Education Conference Proceedings* (pp. 51-57). College Station: Texas A&M University.

Dooley, K., Richards, L., & Lindner, J. (2002). Let's consider the learner. Top 10 course design considerations. *Proceedings of the 9th Annual International Distance Education Conference.* Retrieved March 2, 2004, from *www.cdlr.tamu.edu/dec_2002/Proceedings/Dooley2.pdf*

Hirumi, A. (2003). Get a life: Six tactics for reducing time spent online. In M. Corry, & C.H. Tu (Eds.), *Distance education: What works well.* New York, NY: The Haworth Press, Inc.

Huba, M.E., & Freed, J.E. (2000). *Learner-centered assessment on college campuses: Shifting the focus from teaching to learning.* Boston: Allyn & Bacon.

Kulhavy, R.W., & Wager, W. (1993). Feedback in programmed instruction: Historical context and implications for practice. In J.V. Dempsey, & G.C. Sales (Eds.), *Interactive instruction and feedback* (pp. 2-20). Englewood Cliffs, NJ: Educational Technology.

McMahon, M.J. (2000). Student-centered learning in agriculture and natural resources at Ohio State University. *NACTA Journal, 44*(4), 41-47.

Murphy, T.H., Lindner, J.R., & Kelsey, K.D. (2002, December 11). Authenticated writing competencies of agricultural education graduate students: A comparison of distance and on-campus students. *Proceedings of the 29th National Agricultural Education Research Conference,* Las Vegas, NV. Retrieved January 7, 2005 from *http://aaaeonline.ifas. ufl.edu/NAERC/2002/naercfiles/papers.htm*

The National Commission on Excellence in Education. (1984). *A nation at risk: The full accounts.* Cambridge, MA: USA Research.

Paulson, F.L., Paulson, P.R., & Meyer, C.A. (1991). What makes a portfolio a portfolio? *Educational Leadership, 48*(5), 60-63.

Shale, D., & Garrison, D.R. (1990). Education and communication. In D.R. Garrison & D. Shale (Eds.), *Education at a Distance* (pp. 23-39). Malabar, FL: Robert E. Krieger.

Smith, P.L., & Ragan, T.J. (1999). *Instructional design* (2nd ed.). Upper Saddle River, NJ: Prentice-Hall.

Section IV

Technology Knowledge and Skills

Part IV of this book addresses technology knowledge and skills. An understanding of the technology currently being used in the distance education environment is of particular importance in order to understand design and access issues. Technological infrastructure and delivery strategies, including a variety of multimedia tools to stimulate interactions, are discussed.

Chapter X

Delivery Technology

with
Walt Magnussen, Texas A&M University, USA

 Making Connections

It is our belief that everyone should understand the basic technical lingo of the profession, so this chapter will give you an overview of the tools of the trade. A major consideration when developing a new course or program at a distance is the selection of the delivery technology. You may even have to decide upon specifications for equipment and modify existing space. In the previous section of the book, we explored the principles of instructional design needed to be successful in distance education. Now we will address technology knowledge and skills. You will gain an understanding of the technology being used in the instructional environment, as well as the server capacity for storing data. Of particular importance is the need for instructors and designers to understand that learners have variable access to and abilities with the delivery technologies. Moreover, instruction must be designed for multiplatform use and for future technology development. How do you select or design the learning interface? What combination of technologies and delivery strategies promote engagement and interaction? What are issues with bandwidth and access that impact both instructors and learners?

Introduction

We have previously discussed instructional design and adult learning principles; what you need to know at this point is which delivery technologies will be most appropriate for the design. Some instructors, and usually always vendors, advocate one delivery strategy. We believe that the use of a variety of delivery strategies results in deeper and more meaningful learning. While not based on research, Dale's (1969) Cone of Experience (Figure 1) describes a model for the level of abstractness of various audiovisual media. This model is widely used and provides a nice framework from which to consider delivery strategies appropriate for distance education. Although some researchers have reported percentages of learner retention based upon the type of media, Dale did not offer such recommendations. More direct and purposeful experiences should be used to develop psychomotor skills and more abstract and textural material to enhance higher cognitive development. We also purport that learners have diverse learning needs and preferences, and that the use of blended technologies will improve the ability to reach more learners.

The history of distance learning as a delivery vehicle for formal education has it roots in correspondence study. It later evolved to radio signals and prerecorded media. Later, the information age brought forth the technology tools that we know and use in distance education. In this chapter, we will discuss the following tools: print, audio, audio and video, and computer. We will complete this chapter with a discussion of the Internet and World Wide Web with a description of Web tools for online instruction.

 Internet Connection
http://www.work-learning.com/chigraph.htm

For an interesting discussion on Dale's Cone of Experience, see the Work Learning Research page.

Figure 1. Dale's Cone of Experience (1969)

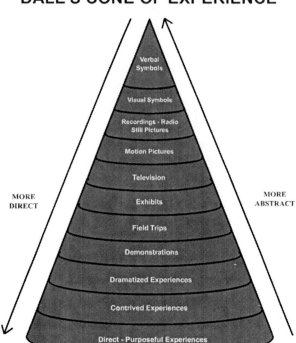

Print media are the simplest and easiest form of distance education to use and have the longest history worldwide. Most print media are transmitted by some form of mail or postal service. In some cases, electronic mail or e-mail is used. Asynchronous communication between learner and instructor is the primary type of communication. Facsimiles or e-mail can be used for delayed interaction between instructor and learner. Chat or instant messaging and phone conversations can provide some "real-time" interaction.

Most print forms of distance learning are batch based. This means that lessons are sent to the learner and new lessons are not sent until the learner has mastered or received some grade or evaluation for the first set. In an electronic format, the second set will not become available to the learner on the screen until the learner has successfully completed the previous lesson. This form of "correspondence" study is still widely used as a cost-efficient and readily accessible form of distance education.

Audio Tools

Audio tools include radio, audio cassettes, videotapes using the audio track only, the telephone, and audio-only conferences. These tools require the smallest bandwidth to deliver electronically. In developing countries, radio continues to be the most frequently used channel to deliver credit courses. In chapter I, we learned that Anadolu University was the largest provider of distance education worldwide, reaching over 500,000 distance learners. The Universidad Nacional de Educación a Distancia in Spain had an enrollment of about 130,000. Many of these learners complete their studies using audio forms of distance education.

Audio conference calls are readily used to encourage dialogue and discourse in distance education courses. Some audio conferences may be as simple as a speaker telephone instrument or as complex as an audio bridge. Because this is a synchronous technology, time zones must be considered. Long-distance telephone calls can also be costly to the individual or institution depending on the infrastructure and support available.

Audio and Video Tools

In this section, we will describe prerecorded media, one-way live video, one-way live video with two-way audio, two-way live audio, and two-way live video. In the discussion of the two-way live audio and two-way live video, we will cover desktop videoconferencing as well as interactive videoconferencing (ITV).

The uses of prerecorded media are really an evolution of print and audio. Some trainers and instructors have moved from audio with graphics to prerecorded media because it provides them with a first-class product of their instructional materials that can be used over and over again. Moreover, most learners have access to equipment that can play prerecorded videotapes.

One-way live video is usually demonstrated by programs downloaded by a satellite receiver to an audience or individual not able to participate by being on location with the speaker or instructor. In this case, the learner would participate as a passive listener similar to watching a television program. Many training programs for organizations with large distribution areas are handled by one-way live video programs. This allows the originating point to distribute information to many sites at one time. The real disadvantage to this tool is the lack of interaction from participants.

To solve this disadvantage of one-way live video, downlink satellite sessions typically allow for a telephone call via a toll-free number. One-way live video and two-way live audio is an answer to the issue of no interaction from the participants. A Web or e-mail/chat feature can also be added, such as with votes for reality TV programs. To restrict the coverage area of the program, one could use the downlink and supplement it with a conference call or use an audio bridge.

Two-way audio and two-way video, also called "compressed video" or "ITV," was a popular delivery strategy in the 1990s. Compressed video allows for the transmission of synchronous video and audio between two or more sites. A computer called a codec compresses the video and audio signals so that they can be transmitted using lower bandwidth. Such technology offers simplicity and affordability. Compressed video can be, and often is, augmented with multimedia (Figure 2).

 **Internet Connection**
<http://www.uwex.edu/ics/cv/about.htm>

This link will take you to the University of Wisconsin-Extension and its Instructional Communication Systems. A complete explanation of compressed video is included.

Figure 2. Diagram of audio and video data compression and decompression

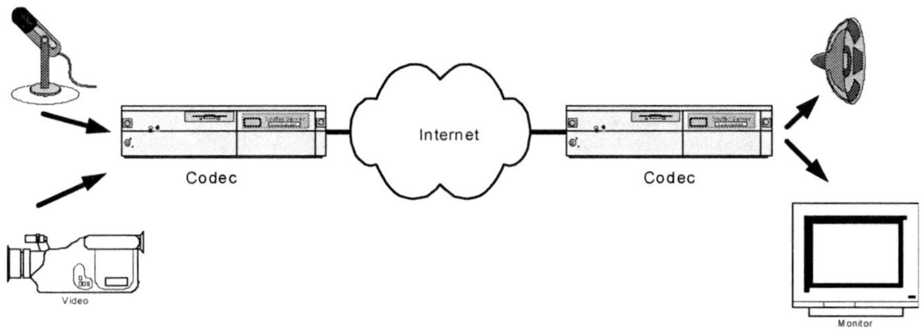

Audio and Video Data Compression and Decompression

In Figure 2, the audio generated by the microphone and the video generated by the camera are analog signals. The signals required by the microphone and the monitor at the remote site are also analog. However, the Internet is not designed to carry analog signals, so the information must be digitized and packaged before the Internet can accept it and transport it. The network device that coverts the information from an analog signal to a digital signal and encapsulates the signal into IP packets is referred to as the "codec" or "coder/decoder" (Solari, 1997).

The first step in the process of converting information is the sampling and quantifying of the analog signal. The second step is to convert the quantified signal into a digital signal. The signal is converted to a data stream by comparing the sampled information to a reference point in a codeword table, which will result in a binary data stream.

Because audio and video can require large amounts of bandwidth (64Kbps for audio and 12Mpbs for video), the data streams are compressed to conserve network resources (Daly & Hansell, 1999). There are several codecs that are used for both voice and video that offer differing levels of compression and provide corresponding levels of quality. In general, the more a signal is compressed, the lower the quality of the received audio and video.

The International Telecommunications Union (ITU), in its H.320 and H.323 specifications, defines the codecs most commonly used for symmetric interactions. The H.320 specification is defined to use dedicated or dial-up Integrated

Services Digital Network (ISDN) lines. The H.323 specification is defined to use the Internet Protocol (IP) as the transport.

Specifications of H.320 and H.323 Standards

Compressed video or ITV became available for classrooms and conferences in the early 1990s at a cost of approximately US$100,000 for a small- to medium-sized classroom. The equipment consisted of a desktop computer, a coder and decoder, television monitors, cameras, and microphones. The main advantage of the videoconferencing classroom was that it could be instructor run, did not require a technician, and could be transmitted via telephone lines. As the technology has evolved, a system can now be purchased for a small classroom that includes a television and a unit that rests on top of the television monitor that includes the coder, decoder, a small television camera, and voice-activated microphones. This unit is quite affordable.

Desktop videoconferencing began with small eyeball cameras mounted on top of the computer monitor and shareware software called "See-You, See-Me." The software was free and the camera was very inexpensive and allowed two computers to connect and share audio and video files. To add additional computers required access to a reflector site. These conferences were very low quality but were available to the masses at a fraction of the cost of videoconferencing classrooms. Units that rest on top of the monitor and plug into the computer now provide desktop videoconferencing using a variety of settings. They can be either analog, high speed, medium speed, or low speed.

The analog videoconferencing uses twisted pair wire. The advantage is that the quality is extremely high because the units are hardwired together. The disadvantage is that the cost increases as you increase physical proximity. These units are usually found in one building or complex.

High-speed videoconferencing is a compressed signal that sends video at millions of bits per second. The algorithm for the compression usually uses MPEG or JPEG and the bandwidth protocol is usually H.310. The normal use for this high-speed desktop videoconferencing is usually in companies or schools.

Medium-speed videoconferencing is the most prevalent form of desktop videoconferencing. The video is compressed so that it can be accessed using the Internet. The bandwidth standard for this type of videoconferencing is

H.323. The H.323 protocol is gaining wider acceptance because the cost of the transmission is free to both users, after paying for the presence on the Internet.

Low speed is videoconferencing at speeds attainable through a dial-up modem (56K–28.8 kps). Here again, the Internet is the ramp used with the H.323 protocol. Due to the slow speed, the quality of the signal is poor and the signal breaks up and freezes frequently. The size of the video screen is limited to a picture-in-a-picture box on the computer screen.

Computer Tools

Computer tools provide great flexibility and accessibility for distance education applications. Covered in this section will be the Internet, networks (LANs, MANs, and WANs), and the World Wide Web.

The Internet

The Internet is the result of a U.S. Department of Defense research project that was concerned about a lack of high-powered computing resources (Comer, 1999). In the 1960s, the Advance Research Project Agency (ARPA) funded a project to investigate ways to make powerful, expensive computers available to more researchers through the use of computer networks. One result of this project was the development of a set of communications protocols known individually as the Transmission Control Protocol (TCP) and the Internet Protocol (IP), known together as TCP/IP. TCP/IP became a common international networking protocol that was eventually embraced by virtually every computer manufacturer in the world, and therefore, the first crucial step in the creation of the Internet (Cameron, 1999).

The research continued through the 1970s and into the 1980s. Then universities began to widely deploy the Internet technologies for their own internal networks. In the 1990s, the same network architectures that had been used by higher education during the previous decade were deployed by the private sector, and the Internet soon became a huge commercial success (Underdahl & Underdahl, 2000). As was defined in chapter I, the Internet is a large network of interconnected computer networks that share a common language.

While the exact size of the Internet is impossible to measure, some estimates show its growth starting slowly in 1982, reaching over 1 million connections by 1993, and skyrocketing to over 30 million connections worldwide by 1996 (Comer, 1999). Nielsen/NetRatings (2004) reported that over 106 million people accessed the Internet in 2003. Moreover, Internet users spent an average of 3 hours and 37 minutes per month using Internet applications.

The Internet has evolved into a best-effort delivery system (Black, 2000). This means that the network (Internet) makes no guarantee that the packets will be received in the order in which they were sent, or that the length of time needed to deliver each packet is appropriate for the type of packet being sent, or that the packets will even be delivered at all. While there is no guarantee of proper packet delivery, the Internet is designed to minimize loss and delay due to traffic congestion, hence, the term "best effort" (Hall, 2000).

Most early applications on the Internet were not negatively impacted by this best-effort system (Durham & Yvatkar, 1999). The TCP/IP protocol is designed to be forgiving in that it will ask for retransmissions of lost packets, is not sensitive to variations in time of delivery, and will reorder the packets into the appropriate sequence if they are delivered out of order. Applications that rely on best effort include e-mail, Web browsing, file transfer, interactive login, and others (Comer, 1999).

While most of these early Internet applications are insensitive to network problems, a new breed of applications, referred to as "real-time" applications, do not share this attribute. Real-time applications include applications that are symmetric in nature; that is, they have a participant on each end of the network communicating in real time. These applications include voice over IP (H.323 and Session Initiated Protocol, or SIP) and video over IP, which includes H.323 and broadband such as MPEG (Chen, 1996).

LANs, MANs, and WANs

Local Area Networks (LANs), Municipal Area Networks (MANs), and Wide Area Networks (WANs) are the building blocks on which all applications are delivered. LANs are typically built within a building or contiguous buildings and run at high to very high speeds (billions of bits per second up to 10 billion bits per second). MANs are typically fiber optic or wireless-based networks that connect two or more buildings within a community to each other, also at high and very high speed. WANs connect LANs and MANs to each other over much longer distances, usually using circuits provided by carriers. These circuits can be carried over fiber optic cable, microwave, wireless, or satellite facilities. Typical speeds for WAN connections include 1.544 million bits per second (T1 speed) to 155 million bits per second (OC-3 speed).

While LANs, MANs, and WANs are built out of the same type of equipment (routers) and the same Internet Protocol (TCP/IP) as is the Internet, the difference between them and the Internet is that the organization controls traffic carried on them. This means that the organization running the network can prioritize critical traffic and rate shape or limit less critical traffic during periods of congestion. The result is that LANs, MANs, and WANs can be built to provide better than the best-effort delivery system that is provided by the Internet.

LAN technology first appeared in the early 1980s. Four initial technologies were ARCnet, Appletalk, ethernet, and Token Ring. These four architectures were somewhat proprietary to Novell, Apple, Digital Equipment Corporation, and IBM, respectively. In the mid-1980s, ethernet was formally adopted as the IEEE 802.3 standard and Token Ring became the IEEE 802.5 standard. The other two protocols were never formalized into standards and have since disappeared. Today the IEEE 802.3 ethernet protocol has become the leader and is the foundation of the vast majority of all networks.

Ethernet is a LAN architecture that is designed to connect all computers and servers within the building or campus on one data highway capable of delivering all applications. All traffic is broken down into a unit called the packet for delivery to the proper destination by the ethernet network. A message such as an e-mail or Web browsing request is usually made up of several packets sent one at a time. The ethernet packet or frame has a standard format allowing all vendors to interoperate with each other (Comer, 1999). The following is the frame format as defined by the IEEE:

- **Preamble:** Provides synchronization and indication of the beginning of a frame.
- **Destination Address:** Provides the unique six-byte ethernet address of the device that is supposed to receive the packet. This address can be a computer, server, printer, telephone (in the case of voice over IP or VoIP) if the final destination is on the same LAN or the default router if the final destination is on some other LAN.

- **Source Address:** This is the unique six-byte ethernet address of the device that originated the packet or frame.

- **Frame Type or Length:** The section of the packet specifies either the type of data that is contained within the frame or the length of the frame.

- **Payload:** This is the actual data that the application is wanting to be delivered. It could be considered the letter inside of the envelope. This section of the packet can vary between 46 and 1,500 bytes in length.

- **Cyclical Redundancy Check (CRC):** This is four bytes of error detection information that is used to verify the integrity of the data. The data are run through a CRC algorithm as it is received and the calculated number is compared to the CRC number calculated at the source device and sent in this part of the frame. If the received CRC bytes do not match the calculated CRC bytes, the frame is flagged as an erred frame and the application will usually ask for the frame to be retransmitted.

LANs are made up of two types of devices: the first could be considered infrastructure devices and the second network servers. The infrastructure devices consist of the network interface cards that connect the hosts to the network, the wire plant, the switches, and the routers. The servers each provide a specific network function such as file or print serving, domain name resolution, firewall, rate shaping, network management, and address assignment. The following are the LAN infrastructure devices:

- **Network Interface Cards (NICs):** The NIC is installed in each host (computer, printer, server, or router). Connection speeds are usually 10 million bits per second or 100 million bits per second. Most NICs support both speeds and will select the appropriate speed utilizing an autonegotiate process. Each NIC has a unique 12-byte address that is used to deliver the appropriate information to the appropriate computer.

- **Wire Plant:** The wire plant typically consists of four-pair, twisted-pair wires between the offices or classrooms and the wiring closets. Wiring closets are typically connected to each other using fiber optic cable. According to the IEEE standards for twisted pair ethernet, the wire between the host and switch cannot exceed 100 meters (about 320 feet). This results in the requirement for several wiring closets in most schools or campus buildings. The fiber optic cable installed between the wiring closets will support ethernet connections at distances of up to 550 meters for 1-billion-bit-per-second connections and up to 2,500 meters at lower speeds.

- **Hub or Switch:** The hub or switch is the first network aggregation device closest to the end user's computer. It typically supports 24 computers, servers, or printers at 10, 100, or 1,000 million bits per second. The hub is a nonintelligent device that sends all information input into any one port or to all other ports. The switch looks at the NIC address of each computer and only sends the information out to the port with the appropriate destination. The net difference is that switches are better able to minimize traffic congestion that would adversely impact applications. In most newer networks, the hubs and switches are connected to routers with gigabit connections (billion bits per second).

- **Routers:** Switches and hubs simply look at the NIC addresses when making a forwarding decision. They have no understanding of the Internet Protocol (IP) that is used to deliver the information globally. As a result, they are only used in the local area network (LAN) environment, which is typically the building, campus, or school district. The hubs and switches will deliver information to computers that are on the same LAN and the default router if they are not. Routers in turn make their forwarding decisions based on the IP addresses. Routers typically connect to each other within the LAN at high speeds (gigabit) and connect to the Internet at speeds from 1.5 million bits per second (T1 circuit) up to a billion bits per second (gigabit). While the T1 speed connections have been in use for several decades, the gigabit connections have only been available since about the year 2000. They have been made possible through the deployment, by service providers, of fiber optic-based Metropolitan Optical Networks or MANs. The following are the MAN or network server devices:

- **Firewalls:** A firewall is a specialized computer that is installed between the institutional LAN and the Internet. Depending on its configuration, it can be used to filter inappropriate information such as sexually explicit Web pages. It can be used to protect the institutional LAN from viruses and other harmful traffic, and it can be used to prevent computers from within the institution from being used to launch attacks on the rest of the Internet were a virus to infiltrate the security. Firewalls are policy-based devices and the policy must be consistent with values, missions, and goals of the institution that it is protecting.

- **Domain Name Servers:** The Internet Protocol uses addresses that are unique and globally assigned. The IP addresses are represented by a four-byte address such as 128.194.15.2. Blocks of network addresses are assigned to institutions to be reassigned to individual computers. To prevent end users from having to remember these complex numbers, a domain name system is used. A domain name such as "mycomputer.myuniversity.edu" is mapped to the unique IP address. Each institution that is assigned addressing authority over a block of numbers is required to maintain a Domain Name Server (DNS). Each time anyone on the Internet attempts to communicate with anybody else, the process begins with DNS servers communicating with each other until the name is resolved to a number. Due to the critical nature of this process, the DNS servers are replicated with a backup server located at some other institution. The location of backup servers is an informal process which is typically negotiated between any two networks on the Internet. The DNS process is a hierarchical system with several root servers strategically placed on the Internet that point name servers in one domain to the authoritative name server in another domain.

- **Rate-Shaping Devices:** These devices, also referred to as "packet shapers," are relatively new. They exist as a response to the explosive demand in bandwidth that came about in the early 2000s as a result of peer-to-peer applications such as music sharing. They establish several (usually eight) queues for different types of traffic, ensuring delivery of higher priority traffic first. This process enables delivery of traffic that is sensitive to network impairments such at packet loss. One important attribute of the Internet is that it is a best-effort delivery system (discussed earlier in this chapter). Network devices have the right to discard traffic that exceeds their capacity to deliver. Most applications compensate for this by retransmitting lost information. Because some applications are more sensitive to packet loss, the traffic-shaping devices can be configured to discard types of traffic where the impact is not as significant.

- **Network Management Servers:** These are devices that monitor the performance of network devices, report changes in operational status of a device, and support security, such as end-user authentication. These processes will usually involve several separate servers, one for each function, often located in a Network Operations Center (NOC). The performance monitoring systems typically report bandwidth utilization in terms of percentage used. The Network Management System (NMS) polls critical devices on a periodic basis and reports an event to the NOC if a device fails to respond, or reports an operational status that does not exceed a preset threshold. These NMS servers communicate to the individual network devices using an IP protocol called the Simple Network Management Protocol (SNMP). Authentication servers are used to centralize lists of authorized users in a campus environment. This is often necessary as there are countless applications on many servers in most institutions. New users can be easily added and users who no longer require access can be quickly and efficiently removed.

- **E-mail Servers:** E-mail servers are repositories of incoming and outgoing e-mail messages. E-mail servers will accept incoming messages and hold them until the recipient is ready to retrieve them. This is necessary because most end users' computers are turned off at least a part of the day. The sender would have to repeat continuously transmission attempts until the recipient's computer is turned on if it were not for a centralized server. These e-mail servers also filter traffic, discarding messages with viruses attached and unsolicited messages such as spam mail.

The World Wide Web

Back in Chapter 1 we delineated the difference between the Internet and World Wide Web (WWW). The WWW is a way of accessing information over the Internet using a browser to access documents via hyperlinks. In response to the demand for education and training, more courses and programs are being developed and offered using Web-based instruction. Web-based instruction is becoming more commonplace due to the increased "proficiency in basic Internet skills, and shrinking barriers with respect to accessing and using the Internet" (Lindner, 1999, p. 37). Allen and Seaman (2003) reported that during fall of 2002, more than 1.6 million students took one or more online and/or blended/hybrid courses. They further projected an approximate 20% increase in the number of students taking online and/or blended/hybrid courses during fall of 2003. Allen and Seaman operationally defined online learning into one of three categories. Web facilitated courses have 1% to 29% of course content delivered online. Blended/Hybrid courses have 30% to 79% of course content delivered online. Online courses have 80% to 100% if course content delivered online.

Table 1. Type of course, percent, and description for Web instruction

Type of Course	Percent of Course Content Delivered Online	Description
Web-developed	80% to 100%	Courses delivered entirely on the Web.
Web-dependent	30% to 79%	Courses that have major content components on the Web, but rely also on other delivery methods.
Web-supported	1% to 29%	Courses that have auxiliary materials, links, additional readings, and support materials on the Web.
Adapted from Murphy & Karasek (1999) and Allen & Seaman (2003).		

Murphy and Karasek (1999) stated that Web-based instruction can be classified into three categories: Developed, dependent, or supported. Courses delivered entirely on the Web are called fully developed. Web-dependent courses have major content components on the Web, but rely also on other delivery methods. Courses that have auxiliary materials, links, additional readings, and support materials on the Web are referred to as Web-supported (Table 1).

Web-based instruction can promote interaction among and between learners, instructors, and content with such features as discussion groups, email, chat, instant messaging, streaming media/video, animations, application sharing, and IP audio/video conferencing (Olliges, Wernet, & Delicath, 1999). A study by Lindner and Murphy (2001) found that learners perceived that Web tools contributed to their ability to accomplish the measured objectives of the course. This perception, however, was diminished when students did not have easy access to reliable computers with high-speed connections to the Internet. Freeman (2000) found that lack of reliable access, both at home and on-campus, to computers and the Internet was a major barrier that negatively affected students' ability to participate in Web-based courses. Providing minimum recommended system specifications for learners considering enrollment in a Web-based courses helps learners make informed decisions about whether or not to enroll.

Web-based instruction can create student-centered online learning environments (Web-based Education Commission, 2000) by providing access to instructional materials and feedback on progress in the course. Instructors and trainers should not assume that learners prefer to receive course materials exclusively from online sources and should make course materials available through a variety of channels. As more instructors make course materials available online, additional inquiry is needed to assess the advantages and disadvantages of such practices (Heinich, Molenda, Russell, & Smaldino, 1999).

As learners become more familiar with Web-based courses, it is expected that learning, instructor effectiveness, and course efficiencies will improve.

Conclusion

Distance education delivery technologies have evolved over the years. Print and audio tools, compressed and desktop videoconferencing, computer tools, and combinations of all of these tools provide the delivery technologies or communication tools needed to bridge the distance between learners and instructors. Educational technologies can contribute to learning the stated objectives of a course, can be neutral, or can distract from learning. Although the decisions made at the instructional design stage determine the delivery strategies needed to promote interaction and engagement, a widely supported position is that technology can contribute to learner success (NCES, 1999).

As noted in previous chapters, "motivated students learn from any medium if it is completely used and adapted to their needs. Within its physical limits, any medium can perform any educational task. Whether a learner learns more from one medium than from another is at least as likely to depend on how the medium is used as on what medium is used" (Heinich et al., 1996, p. 27).

What technology tools will be needed to create an interactive learning environment? Consider Table 2 for selecting media and methods in distance education to help you with that choice. This relates back to your understanding of instructional design and will be useful in completing the application exercise.

Table 2. Selecting media and methods in distance education

	Information	Procedures	Principles and Concepts	Attitudes and Values
Audio	Readings Audio Videotape Lecture Learner presentation Guest speaker	Demonstration Lecture Readings	Class discussion Peer teaching Case studies Panel discussions Group projects	Reaction panel Debates Panel discussions Class discussions Case studies Role playing
Audio Graphics	Readings Audio Videotape Lecture Learner presentation Guest speaker	Demonstration Lecture Readings	Class discussion Peer teaching Case studies Panel discussions Group projects	Reaction panel Debates Panel discussions Class discussions Case studies Role playing
Two-Way Audio One-Way Video	Readings Audio Videotape Lecture Learner presentation Guest speaker	Demonstration Lecture Readings	Class discussion Peer teaching Case studies Panel discussions Group projects	Reaction panel Debates Panel discussions Class discussions Case studies Role playing
Two-Way Audio Two-Way Video	Readings Audio Videotape Lecture Learner presentation Guest speaker	Demonstration Lecture Readings	Class discussion Peer teaching Case studies Panel discussions Group projects	Reaction panel Debates Panel discussions Class discussions Case studies Role playing
Computer Conferencing	Reading Guest contributors	Readings Tutorials	Class discussions Panel discussions Group projects	Reaction paper Class discussions Debates Role playing

Whether using the postal service, radio, audio conferences, one-way video, interactive (compressed) video, or the Internet/WWW, these networks and technologies provide the communication channels necessary for the instructional strategy. In the next chapter, we will explore a variety of multimedia tools that can also enhance the learner–interface interactions to promote learning at a distance.

Application Exercise

There are four major distance delivery systems and numerous combinations. Compare the advantages and disadvantages of each of the individual delivery systems to help you select which ones you will use for your lesson.

	Print	**Audio**	**Video**	**Computing**
Cost				
Learner Control				
Interaction				
Flexibility				

References

Allen, I.E., & Seaman, J. (2003). *Sizing the opportunity: The quality and extent of online education in the United States, 2002 and 2003.* Needham, MA: Sloan-C.

Black, U.D. (2000). *Voice Over IP.* Upper Saddle River, NJ: Prentice Hall.

Cameron, D. (1999). *Internet2: The future of the Internet and other next-generation initiatives.* Charleston, SC: Computer Technology Research Corp.

Chen, B. (1996). *Real-time communications in homogeneous and heterogeneous computer networks.* Unpublished doctoral dissertation, Texas A&M University, College Station.

Comer, D. (1999). *Computer networks and Internets* (2nd ed.). Englewood Cliffs, NJ: Prentice Hall.

Dale, E. (1969). *Audio-visual methods in teaching.* New York: Dryden.

Daly, E.A., & Hansell, K.J. (1999). *Visual telephony.* Boston: Artech House.

Durham, D., & Yvatkar, R. (1999). *Inside the Internet's resource reservation protocol: Foundations for quality service.* New York: Wiley.

Farr, C., & Shaeffer, J. (1993). Matching media, methods, and objectives in distance education. *Educational Technology,* July, 52-55.

Freeman, W. (2000). *And now a word from our students.* Paper presented at the WebCT Conference, St. Louis, MO. Retrieved November 7, 2001, from *www.ryerson.ca/dmp/teaching/webct2000.html*

Hall, E. (2000). *Internet core protocols: The definitive guide.* Cambridge, MA: O'Reilly.

Heinich, R., Molenda, M., Russell, J., & Smaldino, S. (1999). *Instructional media and technologies for learning.* Upper Saddle River, NJ: Prentice Hall.

Lindner, J.R. (1999). Usage and impact of the Internet for Appalachian chambers of commerce. *Journal of Applied Communications, 83*(1), 42-52.

Lindner, J.R., Dooley, K.E., & Murphy, T.H. (2001). Differences in competencies between doctoral students on-campus and at a distance. *American Journal of Distance Education, 15*(2), 25-40.

Lindner, J.R., & Murphy, T.H. (2001). Student perceptions of WebCT in a Web supported instructional environment: Distance education technologies for the classroom. *Journal of Applied Communications, 85*(4), 36-47.

Murphy, T.H., & Karasek, J. (1999). Agricultural student perceptions of the value of WWW supported instruction. *Proceedings of the 18th Annual Western Region Agricultural Education Research Conference, 18,* 165-177.

National Center for Education Statistics (NCES). (1999). *Distance education at postsecondary education institutions: 1997–98* (NCES Publication No. 2000-013). Washington, DC: NCES. Retrieved November 7, 2001, from *http://nces.ed.gov/pubs2000/2000013.pdf*

Olliges, R., Wernet, S.P., & Delicath, T.A. (1999). Using WebCT to educate practice professionals. *Proceedings of the Dancing Web Conference.* Retrieved November 7, 2001, from *http://telr.ohio-state.edu/conferences/dancingweb/proceedings/olliges/olligespaper.htm*

Solari, S.J. (1997). *Digital video and audio compression.* New York: McGraw-Hill.

Underdahl, B., & Underdahl, K. (2000). *Internet bible.* Foster City, CA: IDG Books.

Web-Based Education Commission. (2000). *The power of the internet for learning: Moving from promise to practice.* Retrieved November 7, 2001, from *www.ed.gov/offices/AC/WBEC/FinalReport/*

Chapter XI

Multimedia Design

with

Rhonda Blackburn, Texas A&M University, USA, and
Yakut Gazi, Texas A&M University, USA

 Abstract

As you consider developing a course or training program online, a major question is about what multimedia to use, multimedia that will help you accomplish your instructional objectives. What content do you as an instructor want your audience to learn? After answering this question, a course can be developed that achieves the learning objectives and, at the same time, motivates and entices the learners. Understanding how to create material for a course that integrates multimedia is essential knowledge in the planning stages of course development. This integration should be thoughtful with the understanding of how to balance the techniques and tools for optimal learning potential.

Introduction

The term "multimedia" refers to bringing together a number of diverse technologies of visual and audio media for the purpose of communicating. The different multimedia formats include text, graphics, audio, video, animations, and simulations. The use of each technique should have a purpose within the learning objectives of the course. In this chapter, we will discuss the design, use, access, and best practices to consider when using multimedia.

Graphics

Including graphics in your course has one purpose: to deliver complex information in a way that is easier to visualize or understand than words alone. Images can provide further information; include critical information to the content on the Web page; illustrate a concept without the use of confusing numbers or text; create focal points on the Web page to notes, warnings, or important content; represent how concepts or ideas work together; and enhance the interface and help in organizing the page layout.

Designing Graphics

Key guidelines exist for creating graphics to enhance the instructional design of your course. Avoid creating or using large graphics. Depending on the type of medium, you may have to reduce the size of your graphic to 150 x 150 pixels. Your graphic will be distorted if the medium needs to resize the image. If you are designing for television production, you want to have your graphic between 640 x 480 and 800 x 600 pixels. You need to be careful with the bleed area, which is the outer 10% of the screen. This area should not contain any essential or important information as it may be lost with projection. If designing for television production, you also want to maintain a 3 x 4 ratio for image area for non-wide-angle TV screens.

File format is another main element in designing a graphic. The different formats include bitmap (BMP), graphic interchange format (GIF), joint photographic experts group (JPEG or JPG), or portable network graphic (PNG). The medium dictates which file format to choose. For example, if you want to use a graphic for television, you should use the JPG format. When designing for the Web you can use either JPG or GIF. PNG is becoming a standard format, but not all browsers, especially older browsers, support it. On the Web, the file size should be approximately 30k or less. This will allow the page to load at an acceptable rate.

Under what parameters would you choose a JPG or a GIF? GIF is the most widely supported graphic format on the Web; one should use a GIF for a line drawing or if there is a need for transparency or interlacing. When an image has few colors, the GIF format will make the image look sharper than if it were a JPG. JPGs are better as photographs or full-color graphics and when there is a requirement for faster download speed.

Along with size and format, another important design element for graphics is color. The desired design should demonstrate a high contrast between colors and crisp lines. This will allow a screen that is monochrome to convert the graphic as accurately as possible. If designing for the Web, keep the colors to the 216 Web-safe colors. This may not be as important in another medium, but for the desired product to convert nicely between mediums, it is important to use the lowest common denominator. This also works well with television media because the JPG format is usually used to produce the graphic. Be sure not to use only color in the graphic to convey meaning. For example, do not create three balls, one being red, one green, and the other blue, and tell learners that all assignments are denoted with the red ball. If a learner is color blind or using a monochrome screen, colors are not meaningful and the learner will become frustrated with which ball to click.

There are other elements when designing graphics that play a part in the usability and functionality of the graphic in instructional design. It is important to ensure that the design of the graphic is as simple as possible and that the graphics are consistent throughout the course. Without consistency, the learners may have a difficult time determining which graphic is for what purpose. Moreover, when selecting the font type in the graphic, use a sans serif or true font (without feet, such as Arial or Helvetica). While serif (Times New Roman, Courier) works with font size 12 point or higher, the sans serif fonts are clearer and easier to read within a graphic and in settings where there is a lot of light in the room. Also, avoid using all capital letters; when used often, it is quite tiresome for the eyes because you do not see the natural patterns of letter shapes that make for easier reading.

Layout of Graphics in the Course

Knowing when and where to use graphics is essential to developing a course. If the plan is to use graphics for television or videoconferencing, this information should be conveyed to the producer in the script or in the instructor's storyboard. If the Web is to be used to present the course, layout becomes a critical element in the design. Where are graphics used in the layout of the materials? The following tips should be used in designing the pages.

Balance

It is important to maintain a balance between the graphics, text, and white space. Balance refers to how the interface elements are arranged visually, comparing one side of the page to the other. White space refers to the area of the page where no text or graphical element appears. Good interfaces provide ample negative space to help users cognitively differentiate between paragraphs, columns, and text bodies, and interpret the graphical elements easily.

Above the Fold

So that learners do not miss important information you should try to keep the information in the area referred to by the journalistic term "above the fold." In newspapers, the most important information goes above the fold line, so that people see it and buy the paper. Similarly, designers should try to keep the information on a page so that the reader does not have to scroll to see additional information, especially relevant information. This is especially important when designing to higher screen resolutions because individuals with lower resolutions will have to scroll to see it all. And many Web users characteristically are in a hurry—they do not scroll!

Resolutions

When designing on the Web, design to the lowest common denominator. At least 50% of Web users still use 800 x 600 pixel screen resolution; there are still monitors being used that will not go above 800 x 600 pixels. Therefore, it is wise to design accordingly.

Focal Point

Mitchell (2001) explains that after finishing a page, open it up in a browser and stand back. Do not read the text. Close your eyes for a few seconds and then open them. Look at the page. What element on the page first draws your attention? Where does your eye move next? Trace the path that the eye follows when traveling around. If all the elements on the page demand equal attention, then the eye will be confused as to where to go first and where to travel to next. Having a path to important information is essential in delivering a course.

Simplicity

Keep the layout and design simple. Try not to put too much on one page. Crowded screens are difficult to understand and hence are difficult to use. Present one idea at a time. When two or more ideas are on a single page, the main idea to be put across can be confusing to determine.

Consistency

Try to stay consistent with color, text type, text size, icons, and so forth, when developing pages. Designers suggest placing logos, recurring text, buttons, and graphics in a consistent position on all pages. Stick to a color scheme on logos and buttons to have a sense of unity within the site. Keep in mind that users tend to learn and remember locations of information, functions, and controls (NCI, 2002).

Location

Graphics should face the center of the page to keep the attention on the page and not give the appearance of drifting off the screen.

Metaphors

Choose your metaphors carefully as they may not be as intuitive as you think. For example, in the Macintosh environment, Apple uses a trash can as a way to eject a disk. If the designer is used to using a Mac, this may seem intuitive, but to the non-Mac user, dropping a disk in the trashcan means that the user wants to get rid of the information on the disk, and not to remove the disk from the machine. It is important to see how other people see the metaphors used in the icons and images to decide if they are intuitive and understandable.

Access

Designing materials that people can access benefits us all. This idea is called "designing for all" in Europe and in the United States it is called "universal design." The concept of universal design has been used in the field of architecture for generations, yet it has become important in education and training in the past decade with the integration of technology.

Are *all* individuals able to get to *all* information *all* the time? The answer to this question is an overwhelming "NO." Have you ever lost a connection with a videoconference? Have you tried to access materials from the Web using a dial-up modem? If the site is not programmed correctly, you are not able to access the information you need.

Web sites that comply with the World Wide Web Consortium's (W3C) accessibility checklist are more user friendly. The interface is more consistent and easier to navigate. When a Web site is compliant with that checklist, it can also transform from one media or browser to another gracefully. This helps learners who use cascading style sheets to change the Web site interface so that it adjusts to their preferences. Users who have particular interface favorites will learn better and can access materials easier when the materials transform. This also includes learners who choose to use one particular browser over another.

In addition to being more user friendly and compatible, accessible Web sites also allow search engines and screen readers to pick up information from graphics and provide text alternatives to videos. This information is then read by these various software programs, which make the technology work more effectively to provide the reader with the best information possible.

To make sure that a graphic is accessible, there are a couple of HTML coding tips that will help. Use ALT tags to provide a description of images. This helps screen readers and will provide the text while the image is loading. The HTML coding that creates the ALT tag includes alt="*description*" in the following example <img src="*file path and name to graphic*" alt="*description*"

width="*size*" height="*size*">. The description should be short but able to describe or explain the image. For instance, an image of a house that links to the home page the description would be "link to home page," or if the teaching subject is biology and there is a picture of molecules, the ALT tag would include "picture of oxygen molecules."

Another tip that would help when a page is loading is to include the height and width tags in the HTML code. This allows the browser to load the text and all the images at the same time as the height and width tags permit the browser to know where the images will be placed. The coding includes width="25" height="25" in the following example <img src="house.jpg" alt="link to home page" width="25" height="25">. For more information on graphic design, see the Internet Connections below.

 Internet Connections

http://www.mccannas.com/pshop/menu.htm

This site takes you to Photoshop, Corel, Xara, Painter, and Paintshop Pro Tutorials.

http://www.espressographics.com

This link provides resources about print and Web publishing.

http://www.wdvl.com/Graphics

If you are interested in graphics tools, this link provides techniques and examples to guide you.

Audio

Audio has been used in education for many decades from the bull horn to the tape recorder to radio. With the emergence of the Internet as a resource for distance education, audio is used less than video and text (at least in developed countries). There are many instances when a video of a talking head can be replaced by an audio file. The video increases the bandwidth needed to transmit the same material in an audio format.

When should audio be used over video or text? Audio productions give the learners a chance to hear the instructor's voice. The instructor can take advantage of this by showing passion and interest in the subject that is being delivered, which can increase the motivation and interest of the learners. Audio can also provide an opportunity for an instructor to describe or explain a tough concept that would be difficult to convey in print.

In addition, audio can be used to develop a mood for the course. With light music in the background or a depiction of a topic, the instructor can entice the learners into the subject at hand. For example, if an instructor were teaching World History and the topic was war, having an audio file of what a battle would sound like could put the learners right in the middle of the context.

Certain courses lend themselves to audio better than others. A music course can provide the learners with more information through audio. A French course could use audio to help the learners with expression and pronunciation by providing audio clips of words. Seeing the written word many times does not prepare learners to talk fluently in an unfamiliar language. In return, learners could record themselves and submit the audio file to the instructor.

Developing Audio Files

To produce an audio clip, there are certain steps that should to be followed. First, it is important to know the learning objectives and how the audio clips fit into those objectives. Is information being delivered to benefit learning? Where do the clips fit within the course? Storyboarding the materials along with the audio clips can be useful when designing the course.

Make sure that the correct hardware and software is available to record a well-produced audio file. With the software package chosen, it is important to ensure that it can easily record and edit the clips. The most important hardware in which to invest (and usually the most inexpensive) is a good microphone; a good one will retail for around US$30.

Once the correct hardware and software have been obtained, the next step is to create a transcript of what will be said. This helps to keep the clips organized and assists when recording to eliminate gaps and pauses. The intent is not to make the clip sterile, but to cut down on as many "ums" and "uhs" as possible.

Users want to know what to expect. Provide information about the audio clip; a description of the content, the playing length, the size of the file, and how long it will take to download with most common connection speeds. With respect to the length of playing time, avoid making clips longer than seven minutes, if at all possible, unless the clip is organized in such a way that it can be divided into segments up to seven minutes long with the listener doing something in between segments. If plug-ins are required, provide a hyperlink to the site for download-ing the plug-in at the same location as the file learners are attempting to download.

Access

A transcript will not only help in producing the clip, but it will also be beneficial in making the clip accessible to everyone. The transcript is an important aid to individuals who are hearing impaired, learners who are not as good at taking notes as the rest of the class, and for learners who are better with visual rather than auditory learning as previously discussed in chapter IV. As with ALT tags for images, transcripts also assist search engines in locating and cataloging information presented. Remember, it is important to use the concept of universal design. A transcript will benefit many people, not just those with a disability. For more information on developing audio files, see the Internet Connections provided.

 **Internet Connections**

http://www.newhorizons.org/strategies/technology/bratina.htm

To link takes you to an excellent article titled, "Listen up! Using audio files in the curriculum."

http://catalyst.washington.edu/quick/dlaudio_action.html

If you are planning to use audio in your distance learning course, you should check out this link to Catalyst action plans.

http://www.latrobe.edu.au/asianstudies/Buddha/audio.htm

If you would like to view an example of audio in a distance education course, check out this link.

Video

Creating a video of a moving object is essentially creating frequent still images of that object and replaying it. The Catalyst staff at the University of Washington (UWA, 2003) expands this definition by promoting video as a powerful tool for distance learning to help explain complicated concepts and incorporate visuals into the course. It may also allow the instructor to communicate to distance learners almost as if the instructor were there in person. Learners can view the video at their convenience and replay parts that they want to review. However, both pedagogical/andragogical principles and technical requirements must be taken into consideration.

Video can be used to present case studies, hold panel discussions, show laboratory experiments and demonstrations, share field trips, or model behavior. Make sure that the message intended to communicate warrants video and cannot be communicated by other means, such as text or audio. There are still many users on low-bandwidth connections who will appreciate textual information, rather than trying to download a large video file, unless the use of video is necessary for understanding the material. Video files can be saved on a CD and shipped to learners with slower downloading capabilities if this is the case.

Analog vs. Digital Video

Most of us are familiar with video through television and movies. These types of movies are called analog video. Leong (2001) defines analog video as video that is stored using television video signals, film, videotape, or other noncomputer media. Digital video, on the other hand, is the computer counterpart of the analog video.

Digital video has some advantages over analog video. First of all, it is fairly easy to reproduce without losing or degrading quality. It is also important to note that this advantage also brings the issues of copyright and piracy (to be discussed in chapter XII). Similarly, it is much easier to manipulate and reuse the digital video. One other advantage is that digital video requires less bandwidth to deliver through satellite or radio-based transmissions.

Streaming Video

Another important type of video format is streaming video. Streaming videos can be simultaneously played and downloaded over an Internet connection and can be digitized for different bandwidth requirements varying from modem connections to high-speed Internet capabilities. Using streaming technologies allows the person viewing the materials to begin watching a video before the entire file has been downloaded. The designers for Streaming Media World (2001) explain that streaming architectures are made up of codecs. Codecs, short for compressor-decompressor, are basically mathematical formulas for handling your video information (recall the discussion on videoconferencing in chapter XI). As is clear from the naming, they do two things: 1) they mathematically compress video data into something smaller on the sending end; 2) on the receiving end, they decompress that data into some form for displaying the video. This process may be considered similar to the zipping and unzipping of data files.

Because of the poor quality of streaming video, it is often suggested that one digitizes a higher-quality version of the video and make it available for download. The best practice is to make as many different forms as possible to accommodate a variety of needs.

Instead of providing users with low-quality streaming video, always consider a series of high-quality photo images with textual descriptions. This will definitely prove more useful for learners when compared to a video that they cannot see and understand. Although the hardware and software used in streaming determine the quality of the final product, these are some general guidelines to follow before recording or choosing the video segment for your online class.

Creating a Storyboard for the Video

Storyboards are a planning tool to illustrate key concepts in a series of scenes, representing a draft of the presentation. It helps designers envision the video product before they create it. If an instructor is shooting video to use in class, he or she can have, for example, a very simple storyboard depicting what they will shoot, for approximately how long, and what narration will be added with each scene.

Figure 1. Example of a storyboard for creating a video segment (Askew, 2004)

CHEM 101 Topic: Introduction to Lab Equipment	Previous shot: Chemistry Lab – Pipettes
Current Shot: Chemistry Lab – Volumetric Flask Time: 15–20 seconds Action: Show a shot of the volumetric flask and briefly demonstrate its use. Script: The volumetric flask is used to make solutions. It has a precise graduation line in the neck of the flask. A solute is placed into the flask, then the solvent is used to bring the total volume up to the graduation.	
Next shot: Graduated cylinder	

Here is a simple example of a storyboard for this purpose. Instructors can create and customize this storyboard according to their needs (Figure 1).

Creating a Script

The Catalyst staff (UWA, 2003) suggest that planning a script for the video will make the video segment production shorter, more efficient, and eventually a better experience for everyone. After thinking through the video visually (preparing the storyboards), the next step is to produce a script that describes the ideas intended to be taught and the order in which they will be taught.

Try recording on an audiotape cassette player as if explaining the material to the learners. You can use this audio to determine in what order you deliver your materials by creating a transcript. From this transcript, you can then create a script for the actual recording.

Consider using only audio instead of video. For good storytellers who can explain theories and concepts, the audiotape can perform better than expected.

Audio files are considerably smaller to download. The storyboard example in Figure 1 includes a small script as an example.

User Expectations

Provide some information about the video segment your learners are to view. Tell them what it is about, why it is worth viewing, what may be key points for which to look, how long it is, and what they are supposed to have learned once they have watched it. State the file format used if using any type of nonstandard format. Provide links to plug-ins and special programs that users will need for viewing the video.

It is always a good idea to associate the video segment with an activity so that learners can put it in context. The Catalyst staff (UWA, 2003) suggest that instructors should tell the learners two or three of the most important concepts they should learn while watching the video presentation. Unfamiliar vocabulary can also be explained before the learners view the segment. Providing learners with study questions to use before and after viewing will make the video presentation more effective.

Length and User Control

Most online video clips should be brief (less than 10 minutes). We cannot expect online users to sit and watch an hour-long video. For example, instead of putting an entire lecture on the Web as streaming or regular video, consider transcribing the lecture and provide the content of the lecture as a text, Word, or HTML file. This lecture text can be supported by relevant photographs or maybe short clips of any exciting or critical portions of the lecture. A short video clip is an excellent way to gain attention, give an overview of a module, or illustrate the main concepts in a lesson.

In order to give the users the ability to control their learning environment, present the video in segments. Give users the option of choosing from a menu to access to the different portions of a lecture or presentation, especially if the video is not streamed. Streaming video offers limited control where users can fast-forward or rewind, but it still lacks the precise control that can be offered by breaking the video down into smaller and meaningful units.

Display and Version Considerations

The Catalyst staff (UWA, 2003) explain that in a regular face-to-face class-room the instructor has control over the display technologies such as video projectors and overhead projectors. In distance education, however, the visual materials will be limited by unpredictable dimensions of the learner's television and computer screens. When preparing visuals (video, overheads, graphs, and charts), keep in mind that they should display well on smaller screens or those with lower resolution.

For example, when creating a PowerPoint® voice-over and planning to use Camtasia®, you need to reduce your screen resolution to 800 x 600 or 640 x 480 so that the final product with the video/audio controls can be completely displayed on screens with lower resolution. If creating the voice-over video on a 1,024 x 768 screen, monitors with lower resolution will not be able to display it in its entirety, and learners will have to scroll to the sides or up and down to be able to view the video.

Most users take a long time to upgrade to newer versions of the software. Therefore, it is always a good idea to stay one version behind. If it is really necessary to use a cutting-edge technology that will require users to upgrade or download a new software or plug-in, intrigue them with a preview that they can view with the standard or existing technology. Those who are interested can then download a plug-in or upgrade to a newer version.

Access

Moving away from simple text or HTML always introduces the risk of losing some users, users with insufficient technology or users with disabilities. Any-thing that can be done to provide these users with alternative forms of accessing this same information will prove invaluable. People with hearing disabilities can be supported by the use of captions on videos whereas it is trickier to support users who are visually impaired. Traditional ways of supporting these users is to provide textual descriptions of what is in the video, which can be read by a screen reader. However this audio might clash with the audio of the video. So some prefer to provide two different versions of the same video, one with audio for regular users and one without the audio for users with disabilities, or provide the audio separately from the video. For users with reduced vision, support them by using a larger image.

Uses for Videos in Distance Courses

The following uses for videos can be suggested for distance courses, especially for online courses:

- A video with a brief introduction of the instructor and the course.
- Short video introductions/overviews to each unit, week, module, and chapter.
- Videos of short demonstrations, such as a lab procedure in a physics or chemistry course. For example, showing how to put on protective gear properly for a Power Plant Management course, or how to take a pulse in a Nursing course are good uses of video for distance training.
- Interviewing an expert in the field or incorporating a guest lecturer through a video commentary.

For more information on video components, see the Internet Connections below. Now we will examine two other types of multimedia: animations and simulations.

 Internet Connections

http://catalyst.washington.edu/planning/dlvideo_prep.html

If you are planning to produce a video for distance learners, this site will be useful.

http://catalyst.washington.edu/quick/dlvideo.html

This link takes you to the Catalyst action plans on the use of video in distance learning.

Animations

Animation is a visual special effect using progressive images in rapid succession to create the illusion of movement. Cartoons are the most familiar types of animations. Flash® is the most popular program to create animations today.

Animations can be useful as a method for constructing models and displays. They can be interactive and used as simulations of objects and locations. Small-size animations can be created and successfully used on the Web, keeping in mind that there should always be a clear purpose in using them. Most users are annoyed by constantly moving and flashing objects unless they are essential to the idea to be communicated. Although it may be aesthetically more pleasing, animations may not always be worth the time and effort.

Meaningful animations, especially three-dimensional (3-D) ones, can be complex to design and create. Some animation files with high interactivity and detail can be too large for the users. Learners with limited computer capabilities or those with disabilities can find it very difficult to handle the animations.

Usability studies show that looping animations that continue endlessly are usually irritating to users. It is also advisable to open the animations in a new browser window so that users can close it once they are finished viewing. By doing so, a moving object at one corner of the page will not distract them when they are reading the content on some other part of the screen.

Simulations

A simulation is often considered a working model of reality. Typically, a series of photographs, drawings, videos, or sound recordings, simulations create the impression of acting in an artificial world. Simulations prove useful because of the reduced costs and risks of the real experience. They also provide almost real examples of experiences that are impractical or impossible to reenact (e.g., simulation of an atomic bomb or a plane crash). As multimedia computing becomes cheaper, we are beginning to see copies of the real world. Simulations give learners an opportunity to experience the world and interact with it on their own. Instead of clicking buttons, for example, simulations would open doors and deal with real-world objects.

Simulations also provide users with the opportunity to construct their own knowledge by actively engaging with the learning objects and environment. Through computer technologies, these experiences can be tailored according to the needs of the learner. The simulation format used should be based on the nature of the audience, the technology available to them, and the cost of creating the material in this format. Check out the Internet Connections below for more information on the use of multimedia in distance education.

Internet Connections

http://www.webstyleguide.com/multimedia

To learn more about the applications of multimedia, try this Web site.

http://www.useit.com/alertbox/9512.html

This link is a useful resource on using animations and video on the Web.

When making decisions about multimedia design, you should also consider whether it is necessary for the learners to work synchronously (online at the same time when communicating) or asynchronously (online at different times when communicating) with other learners or if they should do the activities individually. Figure 2 illustrates four combinations of synchronous and asynchronous delivery strategies based on time and location.

Figure 2. Delivery strategies based on location and time

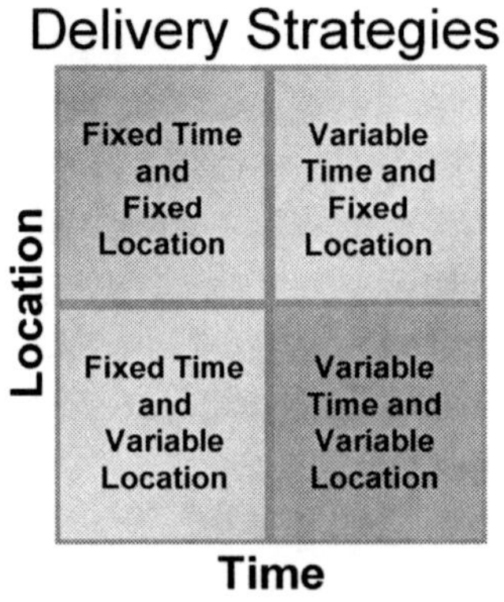
Delivery Strategies
Fixed Time and Fixed Location
Variable Time and Fixed Location
Fixed Time and Variable Location
Variable Time and Variable Location
Location
Time

Conclusion

In this chapter, we explored various types of multimedia and the reasons for using them: (1) Use audio if learners need to hear an explanation of something or if the instructor needs to explain a topic in more detail. (2) Use video if there is a need for the learners to see something that they could not see without the video, such as a demonstration or documentary. (3) Use simulations if it is important to show your learners how to do something and it is necessary for them to interact with the product. (4) Use .html files when there is information that is necessary for the learners to view and read easily on the computer. (5) Use .pdf documents when your presentation needs to stay in the same format as it was developed, for example, a chart or diagram. (6) Use presentation software, such as PowerPoint®, if learners need to be able to print the presentation in note format (this could also be printed to .pdf). (7) Use a spreadsheet software, such as Excel®, if learners need to be able to manipulate the data. (8) Use word processing software, such as Word®, if learners need to collaborate together on one document. (9) Use images if there is a need to show a graphical display of an item.

Consider this question: How much explanation about each software product should be provided to the learners as they are learning or is this to be a self-directed activity? Be careful that *all* learners have access to the software that they are expected to use, for example, Word®, PowerPoint®, Adobe Reader®, Real Media®, or any plug-in. If a plug-in is needed, make sure the learners have the link to download it. Remember, if the learners do not have a technical point of contact, the instructor will need to know the technical aspects of all the technology that is being used, because he or she will be the "go-to person" for problems. The instructor should know the difference between browsers, the different plug-ins, the difference between upload and download, and how to use the technology.

When planning which way to deliver the content, the instructor should have already completed a needs analysis on the course. A needs analysis will include determining technology skill possessed, technology access, age, background, and skill level, to name a few. Another aspect to consider is learning styles. Using a variety of multimedia techniques will accommodate more learning preferences. A visual learner gains knowledge best through text-based files, videos, demos, simulations, images, diagrams, and charts. Auditory learners learn best through audio files and videos that contain audio. Kinesthetic/tactile learners learn best with simulations, interactive tools (discussion boards, chats), and collaborate documents.

Technology changes rapidly. How can an instructor keep up with the latest and greatest in multimedia software? Understanding how to search the Web is an effective method, and evaluating sites for authenticity and accuracy is very important when using technology in the curriculum. It is important to pick a search engine that you like and feel comfortable searching. Clifton, Dougall, Serrano, and Tamas (2001) suggest that when evaluating Web sites, one should consider accuracy, authority, objectivity, currency, and coverage.

It is important to provide the best learning environment. Planning is essential and having clear and concise learning objectives are key components to the planning. Knowing how to deliver content makes the development less cumbersome and more enjoyable. Knowing when to use one type of media and when to use another becomes a development strategy within the instructional planning process. Multimedia can enhance learner motivation and retention if it matches with the learning objectives and course structure. The application exercise will help you to apply your understanding of instructional design and technology knowledge and skills as you build your own lesson for distance delivery.

 Application Exercise

Pick a concept, content, or activity from your lesson that you would like to convert to a multimedia format.
1. List the learning objectives.
2. Decide on the delivery strategy (online, television, videoconferencing, etc.).
3. Choose the multimedia format (audio, video, animations, etc.).
4. Gather or develop the materials needed.
5. Determine the characteristics of the audience (where are they located, what kind of technology do they have, when do they access the material, etc.).
6. Develop a script and storyboard of the multimedia piece.
7. Develop alternative ways to provide to the learners along with the original format you chose.

References

Askew, J. (2004). Science lab equipment. Retrieved February 15, 2004, from *www.howe.k12.ok.us/~jimaskew/labeq.htm*

Clifton, A., Dougall, S., Serrano, S., & Tamas, B. (2001). Course profile: Designing your future. Retrieved February 15, 2004, from *www.curriculum.org/occ/profiles/11/pdf/GWL3OP.pdf*

Leong, C.T. (2001). Video. Retrieved February 15, 2004, from *www.infocomm21.com.sg/showcase/mmb7/tongleong/video.html*

Mitchell, B. (2001). Chapter 1: Design issues in HTML. Retrieved February 15, 2004, from *http://siggraph.org/education/materials/graphics_design/mitchell_S96/chap1_3.htm*

National Cancer Institute (NCI). (2002). Research-based Web design and usability guidelines. Retrieved February 15, 2004, from *http://usability.gov/guidelines/layout.html*

Streaming Media World. (2001). Shooting Video for Streaming (2). Retrieved February 15, 2004, from *http://smw.internet.com/video/tutor/streambasics1/index2.html*

University of Washington (UWA), Catalyst Staff. (2003). Produce a video for distance learners. Retrieved February 15, 2004, from *http://catalyst.washington.edu/catalyst/planning/dlvideo_prep.html*

Section V

Administrative Issues

Part V of the book covers administrative and management issues. We address the major concepts that instructors, learners and administrators need to know when delivering instruction at a distance. These issues cover learner support services, technical support, and copyright awareness and compliance. We will introduce the reader to some general techniques for program evaluation and provide specific techniques of how to apply the tools of program evaluation to the distance education context.

Chapter XII

Course and Program Management

 **Abstract**

Now that we have covered technology knowledge and skills, it is time to review other issues that can impact success in a distance course or program. A major consideration is course or program management. This is an area with a variety of policies and guidelines. Course and program policies, procedures, and management may be coordinated by a centralized unit within an institution, but instructors and instructional designers need to be aware of the applications of management issues. What are the policies and procedures that should be implemented to ensure that best practices are met?

Introduction

Distance training and education is desirable for working adults and others who want flexibility and control over their own learning. The mobility of today's workforce, the needed skill upgrades, and a generation reared on interactive media and various technologies help to promote its acceptance.

Distance learning programs that are highly successful do not just happen overnight; they are a result of careful planning and management. There are many ways to structure management for distance learning programs. This chapter will discuss the keys to success in the planning and implementation of distance programs: conducting a needs assessment, distance learning as a way to revitalize existing programs, using multi-area evaluation, focusing on learning and not the technology, marketing the program, and using technicians. Also, academic policies, fiscal policies and budgeting, faculty policies, copyright and fair use guidelines, and student support issues will be discussed.

Keys to Successful Distance Learning Programs

Chute, Thompson, and Hancock (1999) provide keys to successful management of distance learning courses and programs. These keys, and others we will discuss as well, can be adopted in for-profits, nonprofits, governmental organizations, and the military, both in the United States or in other countries.

Conduct a Needs Assessment

The first key to a successful distance learning program is to determine why the program is needed and who the potential audience is. It is imperative that a proper assessment of needs by potential learners and institutions to be served be completed before the first class or training program is offered. This could include soliciting information from individuals (key informants) whose testimony or description of what exists or what is needed for the client population is available and credible. You could also host a community forum to take the pulse of the community and to garner grassroots support for the training need. Another practice is to develop case studies in which greater in-depth analyses of training needs could be done. Convening a focus group of employers in the field to determine gaps in the preparation of their employees can also help program planners design for a match between content delivery and need of participants. Convening a focus group of employees to determine their needs can also be done, especially, as there exists considerable evidence that often employers and first-line supervisors do not know what are the critical competencies and skills needed by the employees in the organization, competencies that get employees fired if they don't possess them. A survey could be distributed via mail or a Web-based form tied to a database to capture needs of clients. A document analysis of other programs to review their curriculum and delivery strategies can provide valuable insight into whether your proposed program has a unique niche or is a replication of an existing program, especially if evaluations exist on the effectiveness of those other programs. Examining other programs can also provide valuable contacts on potentially sharing course components and logistical responsibilities.

Use Distance Learning as a Way to Revitalize Existing Programs

The second key to success is to explore distance learning as a way to change an on-site program or to revitalize a program for a new audience. This could be a program that is suffering from low enrollment. One reason may be the population in the potential audience might be more technologically savvy and traveling continuously for their jobs; therefore, conversion to distance learning could jump-start the program and provide the necessary flexibility. According to Bates (1997), "the widespread introduction of technology-based teaching will require such fundamental changes to an institution that its use should not be embarked upon lightly, nor will it necessarily lead to any significant cost savings, but nevertheless such an investment will still be necessary if universities [or other training organizations] are to meet the needs of its students and society at large" (p. 1).

Use Multi-Area Evaluation

The third key to success is to use a multi-area evaluation approach. Distance learning should be evaluated in three areas: functional, managerial, and instructional. Functional is related to the technical and design area and is associated with equipment requirements and specifications. The quality of the program's outcomes cannot be achieved in isolation from functional and managerial levels of quality. Managerial is related to how successfully the relationships within and outside the parent organization are fostered and managed as they relate to the distance education mission of the organization. The instructional area is concerned with measures of program outcomes and evaluation. This area also pertains to "the teaching-learning requirements of particular educators and learners and the nature of the subject matter, as well as the parent organization's mission and resources and the external forces that affect all decision making" (Duning, Van Kekerix, & Zaborowski, 1993, p. 198). Evaluation of the instructional area involves evaluating both the process (ways and means) and the product (outcomes) of the instruction as outlined in the CIPP (Context, Input, Process, Product) model of program development formulated by Stufflebeam (1974).

Multi-area evaluation is a fundamental part of the cost justification for distance learning. Part of a multi-area evaluation includes selecting technologies as a part of your strategic planning and decision making. Bates (1997, p. 4) proposes the ACTIONS mode:

Access

Costs

Teaching functions

Interaction and user friendliness

Organizational issues

Novelty

Speed of course development/adaptation

An institution wanting to expand course offerings and access to new clientele can use the ACTIONS model to evaluate decisions about technology choices; interactivity components needed to facilitate virtual communication channels; how trainers and instructors can most effectively transfer the knowledge, skills, and abilities through the technology interface; and how the end user will respond to the technology in terms of novelty and access/availability. More information about evaluating distance education programs will be available in the next chapter based on the principles of good practice.

Focus on Learning and Not on the Technology

The fourth key to success is to keep the focus on learning and not on the technology. The technology continues to change rapidly and personnel responsible for keeping up with the changes can get caught up in the newest toy, forgetting the instructional focus or reason for its application. Unfortunately, this often seems to be the marketing approach for telecommunication companies. For example, the shelf life of most computers is three to five years, but computer companies encourage buying a new computer because of the increases in speed and capability available. Although having a new computer with high-speed access gives the learner some advantages, the decision should not be made on technology alone, but rather on the needs of instruction or learning. Another way to think about this is to consider a computer network. A network is not about the routers, switches, or bandwidth; it is about the content. Without the content, there would be no need for a network. So keep the focus on the content to be delivered to your clients to help them reach their instructional or professional goals.

Market the Program

A fifth key to success is to market your distance learning program, both internally and externally. A nice brochure and intuitive Web site is a good place a start, especially if you have determined the potential audience for your program based on the needs assessment. Use data that can be statistically verified or other methodological formats that are acceptable within your environment. If your clients do not know about your program and success using distance learning, they will not choose it. Use a variety of media to get the word out, such as radio, TV, print ads, personal mail pieces, and so forth. Potential learners will search for programs that meet their program and professional interests and needs. They also want assurance that the program is of high quality. They want to know there is sufficient library, laboratory, or other resources that can be accessed easily and 24 hours a day. They want to know how much it will cost. They want to know who to call for advice on course offerings, scheduling, registration, or other paperwork for their certificate, continuing education, or degree. They will not choose your program unless you deliver a quality program that meets their needs and have proof that you can do so.

Use Technicians

A sixth key to success is faculty support in the development and delivery of the program. It is critical that the trainers or instructors be provided the technical support necessary to create and deliver the courses. Most instructors know their content but do not know how to repair the computers/ITV/networks, and so forth. Technicians can provide support to both the instructors and learners by ensuring that the equipment is functioning properly and the components necessary for the learning community are available. Providing this support for instructors and learners who are not in close proximity to a campus or a corporate setting requires institutional support and planning. A 24-hour help line is useful for learners to get support with technology problems. Many institutions provide virtual tutors using desktop videoconferences, chat features, or simple e-mail. For online instruction, the use of virtual teams or mentors/tutors can alleviate some of the feelings of isolation and provide the social interaction and dialogue necessary for active participation and engagement.

Ensure All Instructors Are Well Trained

It is very important that all instructors are well trained in areas other than knowledge of content. Presentation skills, the use of animation and graphics at a distance, the proper use of audio, audio voice-overs on PowerPoint, use of Web-course tools, and so forth, are examples of critical components of instructor training programs. The focus should be on andragogy, or adult instructional methods, with the technology interface serving as communication channels. The technology should appear seamless and transparent, with the goal being ubiquitous delivery of instruction. The role of the instructor changes in online instruction; thus, instructors should learn how to chunk the content into learning objects and limit massive amounts of text reading and lengthy video downloads or streams online.

Design Programs Specifically for Distance Learning

Careful planning and organization is very important to maintain a balance between human interaction and the use of technology. As an example, the use of projects, papers, and portfolios is an appropriate way to assess learning authentically in adult settings. In some areas, tests and exams may be necessary to measure understanding of the content and confirm competence for continuing education credits or skills-based training. The use of case studies as a reading/reflection/writing exercise is very common in online MBA programs. Such usage is also appropriate in other settings. The main thing to avoid is trying to make a standard face-to-face class lecture transfer as is into the distance education, e-learning, or online setting. Instead, the content must be presented in a different way to be effective.

Use Reliable Equipment

Unreliable equipment or equipment breakdowns tend to be the most common complaint from learners when describing a poor distance education experience. A backup plan should be in place, such as materials being available for downloading off the Web; being able to fax materials to an on-site coordinator for a videoconference so that learners can participate in the activity without ITV; or when the TV signal is lost, use audio conferencing. If you are teaching using a two-way videoconferencing classroom and the video signal is interrupted, you do not have to cancel the class. The audio signal requires much less bandwidth and will more than likely still be active (Chute, Thompson, & Hancock, 1999). The same can be said for online delivery. Network administrators should back up content weekly (if not daily) and provide redundant servers to keep the network functioning.

Policy Issues for Distance Education

It is also important to examine the policy analysis framework for distance education. See Table 1 for a list of key policy areas and issues (adapted from Berge [1998] and Gellman-Danley & Fetzner [1998], as cited by King, Nugent, Russell, Eich, & Lacy, 2000).

There are many different policy areas. You will need to work within the guidelines of your company or institution. For course developers, instructional designers, instructors, and trainers, there are five primary policy areas of

Table 1. Policy analysis framework for distance education

Policy Area	Key Issues
Academic	Calendar, course integrity, transferability, transcripts, student/course evaluation, admission standards, curriculum/course approval, accreditation, class cancellations, course/program/degree availability, recruiting/marketing
Governance/Administration/Fiscal	Tuition rate, technology fee, FTEs, administration cost, state fiscal regulations, tuition disbursement, space, single vs. multiple board oversight, staffing
Faculty	Compensation and workload, development incentives, faculty training, congruence with existing union contracts, class monitoring, faculty support, faculty evaluation
Legal	Intellectual property; faculty, student, and institutional liability
Student Support Services	Advisement, counseling, library access, materials delivery, student training, test proctoring, videotaping, computer accounts, registration, financial aid, labs
Technical	Systems reliability, connectivity/access, hardware/software, setup concerns, infrastructure, technical support (staffing), scheduling, costs
Cultural	Adoption of innovations, acceptance of online/distance teaching, understanding of distance education (what works at a distance), organizational values

importance: academic policies, fiscal policies and budgeting, faculty policies, copyright and fair use, and student support.

Academic Policies

Comments about policies in this section apply more specifically to formal programs being offered through a university system rather than in a corporate setting. Yet some components do apply to both. One consideration is the planning phase of calendaring. Looking throughout the upcoming year and deciding how and when you will accomplish your training goals is an important first step. Some organizations use GANTT or PERT charts, electronic calendars, or paper and pencil calendars; but regardless, dates and formats must be decided. A second area is accreditation of the program through a governing board or agencies such as the USDA Food Safety Inspection Service or a university coordinating board. The program needs the appropriate approvals and credentials. The teams of instructional designers and content experts need to discuss the course infrastructure and interface to ensure quality standards are met. Decisions must be made about course and program evaluation components, grading criteria, admissions, and curriculum review and approval (Simonson & Bauck, 2003). These decisions serve as the springboard for the other policy decisions, such as how will it be funded?

Fiscal Policies and Budgeting for Distance Education

The process of budgeting and accounting used provides a way to track and spend funds that are accountable to the client and institution. Within budgeting, there are two categories: costs/expenditures and income.

The expense side of the ledger can be divided into planning costs, personnel costs, and operating expenditures. Planning costs are usually those costs relating to either start-up of a program or even the planned acquisition of a building for distance education purposes. When planning a facility, one must conduct site surveys and needs analyses, develop PERT charts and critical time lines, and do technology vendor research to determine the direction your organization would like to go.

Regardless of the relationship you might have with a particular vendor, never accept the opinion of one vendor when planning an acquisition. It is unlikely that one vendor will provide you with an unbiased opinion related to a purchase. It is imperative that you find a neutral party to provide you with a technology plan. An example of an entity that can do this type of neutral assessment is Texas A&M University's Center for Distance Learning Research. Vendor research should include a variety of modalities for delivery of material. As you have learned in earlier chapters, not all learners will have the same learning preferences and therefore, not all courses are best delivered using only one form of technology. The previous chapter on multimedia also reinforced this idea.

Personnel costs are costs that vary with the type of personnel required for the job, as well as the type of organization. Some organizations prefer to have an instructional design team on staff to design all the instruction for the instructors/

Internet Connection

http://www.cdlr.tamu.edu

The Center for Distance Learning Research is a training and development center that was created as a joint venture between the Texas A&M University system and Verizon Corporation.

---

trainers. Others have mandated that all their instructors be trained in all the multimedia software and authoring systems so that they can design their own material. The type of organization will dictate the type of personnel that you will need. It is not a good idea to try and train all instructors in all the software needed to design instruction for distance learning. To stay current in all the software is not an efficient use of time for instructional staff.

Personnel costs are hard to estimate due to variation in types of organization offering distance programs. For university or other educational organizations, it is much more economical to hire students than it is to hire permanent workers where salary and fringe benefits are more costly.

Income to purchase and maintain a distance learning system can include contracts and grants, bridging or leveraging other sources of funds, in-kind contributions from other sources, matching funds, or sources from the private sector.

Developing contracts and securing grants is an area that can be challenging. Most grants limit capital purchases for expensive equipment but will fund program areas that promote collaboration between and among institutions. The delivery technology is a by-product of the proposal idea; however, some equipment can be funded to accomplish the delivery goal. In the grants marketplace, this is called "redefining your ideas to find more funding sources."

Bridging or leveraging of funds is a technique that requires thinking outside the box. Bridging is a technique that can only be practiced in the short term and only if the organization with which you are associated has the extra capital to make it work. You may have been notified that you will have budget authority to purchase distance education equipment, either by the notification of a grant award or the newly adopted budget, but this authority will not go into effect for a while. However, the opportunity presents itself now for purchasing distance education equipment. Additionally if you do not purchase it now, you will be subject to a large price increase by the vendor. Bridging is where your host organization agrees to allocate you the funds with the proviso that you will return it to the host account when your account is funded in the future. Many organizations are willing to enter into these types of agreements if doing so will save the organization money in the long run.

Leveraging can occur when you have a source of funds that will not allow the purchase of equipment but will allow the lease of equipment. Leveraging will allow you to request permission to lease the equipment from your host organization. If permission is granted (do not attempt this without the advance permission of the granting organization), you can purchase the equipment and lease it from your host organization. At the end of the lease period, the equipment becomes your unit's property. In any arrangement, make sure you communicate fully with all parties.

In-kind contributions to your distance education equipment needs list occurs when you may take a donation of equipment or the transfer of equipment from one part of the organization to another. In-kind contributions have already been paid in full so the contributing organization is not losing anything and in some cases, can take a tax write-off for the contribution to your organization.

Matching income sources can be difficult to complete unless you already have a source lined up. Essentially, this means that another source has promised to give you a certain amount of money if you can raise a matching amount. For instance, a charitable foundation agrees to give you $100,000 if you can raise another $100,000 in a period of six months. There have been instances where universities have been promised $3 million if they could raise another $3 million for one specific purpose and the university had to return the funds because they were unable to locate the match.

The last commonly considered income source is private companies or corporations. These are usually most effective when soliciting small amounts of income. Local retailers are a good example. For example, some stores will donate to causes that will have an impact on the communities they serve. Some private foundations also support educational initiatives. You may want to check with a grants and contracts office on your university campus or do a library or online search to identify potential funding sources. For a sample budget, see Table 2.

Table 2. Distance education project budget justification

Personnel	Distance Education Project Budget Justification June 1, 2004–May 31, 2005		
	In-Kind	NSF	Total
Project Director, 33%			
Joe Professor			
$5,864/mo. x 3 mo. x 25%		$4,398	$4,398
$6,157/mo. x 9 mo. x 25%		$13,853	$13,853
$5,864/mo. x 3 mo. x 8%	$1,407		$1,407
$6,157/mo. x 9 mo. x 8%	$4,433		$4,433
Principal Investigator, 25%			
Junior Faculty			
$3,573/mo. x 3 mo. x 20%		$2,144	$2,144
$3,645/mo. x 9 mo. x 20%		$6,561	$6,561
$3,573/mo. x 3 mo. x 5%	$536		$536
$3,645/mo. x 9 mo. x 5%	$1,640		$1,640
Instructional Designer, 25%			
To Be Named			
$2,406/mo. x 3 mo. x 20%		$1,444	$1,444
$2,455/mo. x 9 mo. x 20%		$4,419	$4,419
$2,406/mo. x 3 mo. x 5%	$361		$361
$2,455/mo. x 9 mo. x 5%	$1,105		$1,105
Project Assistant			
I.M. Important			
$1,800/mo. x 10 mo. x 75%		$13,500	$13,500
Staff Assistant			
$1,800/mo. x 12 mo.		$21,600	$21,600
Total Salary	**$9,482**	**$67,919**	**$77,401**
Fringe Benefits			
16.75% of SW + $257.16/mo. Health Ins.	$10,633	$20,421	$31,053
Total Salary and Benefits	**$20,115**	**$88,339**	**$108,454**

Faculty Policies

Once the institution or agency has determined the overall academic policies and the fiscal components, then securing distance education equipment and facility renovation can begin. While equipment is being installed, the instructors need to develop the content (often with the assistance of an instructional design team). Policies about compensation and workloads should be clearly stated

Table 2. Distance education project budget justification (cont.)

	Distance Education Project Budget Justification June 1, 2004–May 31, 2005		
	In-Kind	NSF	Total
Travel			
Strategy Retreats Estimated airfare $450/trip x 3 persons 3 days per diem @ $70/day x 3 persons		$1,980	$1,980
Training Program x 3 persons Estimated airfare $450/person 5 days per diem @ $70/day/person Tuition @ $2,000/person		$8,400	$8,400
Presenting Papers to 4 National Meetings for 3 persons San Francisco, Atlanta, Chicago, New York Estimated airfare $500/trip/person 2 days per diem/trip/person x $70/day		$7,680	$7,680
3 visits to the National Science Foundation for 2 persons Estimated airfare $500/trip/person 1 day per diem @ $70/day/person		$3,420	$3,420
Total Travel		**$21,480**	**$21,480**
Equipment			
Video Conference Room			
Dual 32" S-Video Monitor TC System		$35,000	$35,000
Dual RS-449 Cable		$360	$360
Modular Riser 32 in./35 in.		$695	$695
TC2000 1st yr Enhanced Warranty		$900	$900
Student System for TC System		$7,980	$7,980
Cable S-Video 100 ft (3)		$110	$110
Option Wall-Mount Bracket SmartCam		$150	$150
Video distribution amp; 1 to 4; S-video (2)		$200	$200
Table microphones (9)		$500	$500
30' cables for mics (9)		$50	$50
9" color television for table monitor		$350	$350
TV mount for ceiling with yoke (2)		$175	$175
Flange for ceiling		$50	$50
Pipe extension		$25	$25
Subtotal Video Conference Room		*$46,545*	*$46,545*

early as they will be reflected in the budget. If you are involving a university faculty member, will he or she receive a reduction in workload as he or she works on this new distance education project? Are there available design and development incentives such as a mini-grant program or access to multimedia producers or student workers? What kind of training and development is necessary to ensure the skills are matched to the task of creating these instructional materials? How will faculty be recognized for the efforts and

Table 2. Distance education project budget justification (cont.)

	Distance Education Project Budget Justification June 1, 2004–May 31, 2005		
	In-Kind	NSF	Total
Desktop Computers (4 @ $4,672/ea) Pentium® 4 Processor with HT Technology 3.2GHz Extreme Edition Microsoft® Windows® XP Professional 1GB Dual Channel DDR SDRAM at 400MHz 128MB DDR ATI 9800 Pro Graphics Card Sound Blaster Audigy™2 (D) Card w/Dolby 5.1, and IEEE 1394 capability 800GB RAID 0 (2 x 400GB SATA HDDs) Single Drive: 16X DVD-ROM Drive 20.1 in 2001FP Digital Flat Panel A425 Speakers With Subwoofer Wireless Keyboard and Optical Mouse Microsoft® Office Professional 2003			
Subtotal for Computers	*$18,688*		*$18,688*
Total Equipment	**$18,688**	**$46,545**	**$65,233**
Training and Support			
$150/participant x 48 participants		$7,200	$7,200
Technical Support Toll-free telephone number and guaranteed online support for 12 months @ $600/mo		$7,200	$7,200
Total Training and Support		**$14,400**	**$14,400**
Supplies			
Office supplies $200/mo. x 12 mo.		$2,400	$2,400
Curriculum materials $100/mo. x 12 mo.		$1,200	$1,200
Cellular airtime 600 minutes/mo. x $0.25/min. x 12 mo.		$1,800	$1,800
Copying, duplicating, and other costs $200/mo. x 12 mo.		$2,400	$2,400
Total Supplies		**$7,800**	**$7,800**

expertise? Key issues include compensation, design and development incentives, recognition of intellectual property, staff development, and other workload issues (Simonson & Bauck, 2003).

Table 2. Distance education project budget justification (cont.)

	Distance Education Project Budget Justification June 1, 2004–May 31, 2005		
Other Expenses	In-Kind	NSF	Total
Telephone charges, Internet expense $300/mo. x 12 mo.		$3,600	$3,600
Postage costs $50/mo. x 12 mo.		$600	$600
Printing materials for final report 500 copies x $3/copy		$1,500	$1,500
Total Other Expenses		$5,700	$5,700
Total Direct Cost	$38,803	$184,264	$223,067
Total Direct Cost Base	$20,115	$137,719	$157,834
Indirect Cost 45% of Modified Total Direct Costs	$9,052	$61,974	$71,025
TOTAL PROJECT COST	$47,855	$246,238	$294,093

Research has recognized the need for a change and modification of the faculty role in teaching at a distance (Jones, Lindner, Murphy, & Dooley, 2002; Kanuka, Collett, & Caswell, 2002; Miller & Pilcher, 2001). Lack of perceived institutional support (faculty rewards, incentives, training, etc.) for course conversion to distance education formats is one reason why faculty are resistant (O'Quinn & Corry, 2002; Perreault et al., 2002). The ability of an organization to adapt to these changes is influenced by the competence of the staff, the amount of importance the staff places on the role of these technologies to accomplish teaching and learning, and the availability of high quality facilities, equipment, technical support, and training (Dooley & Murphy, 2000).

As indicated by Moore (1997), distance education programs with a commitment to faculty support and training result in higher-quality programs. As the complexity continues and the desire to integrate distance education programs expands, attention must be given to faculty training and support. Enhancing faculty participation requires that resources be directed to provide adequate levels of support and training such that these technologies are used for the benefit of the learners (Howard, Schenk, & Discenza, 2004). It is the

Figure 1. Integration of training, support, and incentives to promote faculty participation

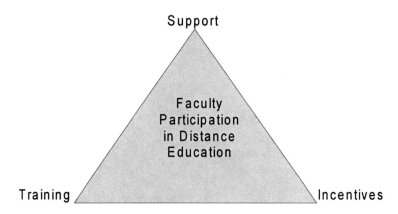

integration of incentives, training, and support that promotes the adoption of distance education delivery strategies by instructors (Figure 1).

Rockwell, Schauer, Fritz, and Marx (1999) find that the primary incentives for faculty participation are *intrinsic* or personal rewards, including the opportunity to provide innovative instruction and to apply new teaching techniques. Other incentives include extending educational opportunities beyond the traditional institutional walls, and release time for faculty preparation. Course designers, instructors, and trainers must also be aware of copyright and fair use guidelines while in the development and delivery stages of a distance program.

Copyright and Fair Use

"If copyright is like property in land, infringement is like moving onto someone's land without permission, chopping down trees, mining coal, and stealing water from the well" (Strong, 1999, p. 178). A brief history of copyright law and relevant provisions and guidelines are provided here to help course developers, trainers, and instructors make decisions regarding instructional design and delivery of programs at a distance.

Copyright provides protection for authors and developers while also allowing flexibility for the public to have access to original works. The first national copyright law originated in the British Parliament in 1710 (255 years after the invention of the printing press), and served as the foundation for current copyright law in England and the United States (Bielefield & Cheeseman, 1997). Copyright law was enacted in the United States in 1790 and typically has been revised about every 50 years—1831, 1870, 1909, 1976 (Bielefield & Cheeseman, 1997; Bruwelheide, 1995). In the 1976 edition, copyright protection was extended to the life of the author plus 50 years. This act also fixed copyright at the moment of creation and did not require the copyright notice for protection (Bielfield & Cheeseman, 1997).

Even though copyright law was being reformed, it was not keeping pace with developments in England and other countries (Bielfield & Cheeseman, 1997). Many nations throughout the world gathered for a conference in 1885 known as the Berne Convention (Berne Union for the Protection of Literary and Artistic Property). The countries represented agreed that they would recognize the copyrights of each other and make any changes to their copyright laws to comply with international standards. The United States was unable to sign because it could not meet the convention's standards. The copyright law revision in 1976 was an effort to join the Berne Convention and thus bring the United States up to international standards. By 1988, the United States had made significant changes in order to sign the Berne Convention (Bielfield & Cheeseman, 1997).

As was mentioned previously, copyright exists to protect the authors *and* to provide reasonable access to the public. Access to the public is expressed through Section 107 of Title 17 of the U.S. Code, called *Limitations on exclusive rights: Fair use*. Four factors determine whether a work may be used: (1) the purpose and character of the use (commercial vs. not-for-profit/educational use); (2) the nature of the copyrighted work (factual vs. entertainment); (3) the amount or portions used; and (4) the effect on the potential market (Bielefield & Cheeseman, 1997; Bruwelheide, 1995; Strong, 1999). A statement "Warning of Copyright" should be included on photocopies or reproductions stating: The original material reproduced here for classroom use is protected by copyright. It is "not to be used for any purpose other than private study, scholarship, or research" (Strong, 1999, pp. 324–325). If you are in the corporate training sector and charge for organization and professional development, then you must secure copyright permission and pay appropriate fees as the fair use guidelines will not apply.

The Commission on New Technological Uses of Copyrighted Works (CONTU) published a document in 1976 to provide interpretation or guidelines for fair use. By the mid-1990s, this work was extended to encompass digital and interactive technologies such as the Internet. The Conference on Fair Use

(CONFU) drafted guidelines for libraries and educational institutions, including multimedia and distance education.

Of greatest concern for online educators are the guidelines for multimedia use. A general understanding based on CONFU is the "10% rule." Ten percent or less of a work can be used without infringement. This rule serves as a guideline, but is not a part of the Copyright Law. It is important to determine (1) how the multimedia will be used, (2) whether there are limitations on the use, such as password protection, (3) if permission can be granted to use copyrighted works, and (4) the appropriate references to credit your sources (Bruwelheide, 1995).

Many people (including course developers, trainers, and instructors) believe that information that can be accessed from the Internet/WWW is in the public domain. This is not true. Copyright protection is available as soon as the material is fixed in a tangible manner and does not require copyright notice. It is not enough to acquire the rights to use materials in a course. The course developer, trainer, or instructor must also notify the copyright holder of the intention to transmit the course over a network.

The Internet makes it much easier to transmit or distribute copyrighted materials. Therefore, Congress enacted new copyright legislation in 1998, the Digital Millennium Copyright Act. The Act has many implications for online education: (1) Infringement liability protections – "service providers" are responsible for ensuring copyright permission is granted before materials are accessible from a Web site; (2) Circumvention of technological protection measures – owners of copyrights should control access and reproduction of protected materials; and (3) Distance education study – permits digital transmission over computer networks and eliminates the "classroom" requirement with provisions for mediated/asynchronous instruction. For distance education study, the service provider should prevent unauthorized access, retain "non-profit" status, and continue to ensure the use of limited portions (Simonson, Smaldino, Albright, & Zvacek, 2003).

Based on some of the changes in interpretation and expansion for distance education in the Digital Millennium Copyright Act, the Senate passed the Technology, Education, and Copyright Harmonization (TEACH) Act in 2001. The TEACH Act emphasizes five changes: (1) It expands categories of works that can be performed in distance education to reasonable and limited portions; (2) It removes the concept of the physical classroom and recognizes that a learner should be able to access the digital content of a course wherever he or she has access to a computer; (3) It allows storage of copyrighted materials on a server to permit asynchronous performances and displays; (4) It permits institutions to digitize works to use in distance education when digital versions do not already exist; and (5) It clarifies that participants in authorized distance education courses and programs are not liable for infringement. Part of the mediated instruction must be under the supervision of an instructor and

Figure 2. Sample letter requesting permission to use copyrighted material (Adapted from Simonson et al., 2003)

Mr. B. Smart
Manager of Permissions
North American Productions
1234 Highly Intelligent Lane
New York, NY 56789

Dear Mr. Smart:

  As a follow-up to my e-mail, I am hereby requesting permission to use your videotape presentation, "The Distance Education Professional," in the distance education course that I will teach this fall. I purchased this videotape from your company for my own personal library in 2002. I think this presentation would be an excellent resource for my students.

  The course is AGED 611, "Advanced Methods in Distance Education," and is expected to enroll approximately 25–30 students, including about 10 participating at off-campus locations. The course will be transmitted to five sites (Houston, Dallas, Austin, El Paso, and Lubbock) via the Trans-Texas Video Network, a statewide telecommunications system. This is a closed-access system, so the transmission will be received only in these five specially equipped classrooms, and no local recordings will be permitted. A backup recording of the entire class period, including the lecture, video presentation, and class discussion will be made at the origination site, and made available to students unable to attend the class. This tape will be provided for viewing only in a controlled access environment with no reproduction equipment accessible. The tape will be erased after two weeks.

  The anticipated date of use is September 17, 2004. I would appreciate your response by September 1, if possible, so that I can make alternative course plans if necessary. I do hope you will honor this request, because your tape is highly informative and engaging, and would be extremely valuable to my students.

  Thank you very much for your consideration. Please contact me directly if I can provide further information.

Sincerely,

Kim E. Dooley, PhD
Associate Professor

requirements for lawfully made or acquired copies of works provide additional safeguards (Gasaway, 2001).

For more information on copyright and the use of copyrighted materials, please use the Internet Connections provided. A sample copyright permission letter is also included (Figure 2). A letter can be sent via e-mail; just make sure that you keep a copy of all correspondence.

 Internet Connections

http://www.utsystem.edu/ogc/intellectualproperty/teachact.htm

The TEACH Act Finally Becomes Law is a comprehensive overview of the TEACH Act and the section 110(2) expanded rights. There is also a checklist at the end.

http://www.lib.ncsu.edu/scc/legislative/teachkit/index.html

The TEACH Toolkit is an online resource for understanding copyright and distance education. This is a joint project of the libraries, Office of Legal Affairs, and distance education at North Carolina State University.

http://www.lib.utsystem.edu/copyright

This site provides a copyright crash course about how ownership of copyrighted materials works, what is fair use, and when and how to get permission to use someone else's materials. There is also a link to copyright law.

http://www.copyright.com

The Copyright Clearance Center (CCC) acts as the agent on behalf of thousands of publishers and authors to simplify the process of obtaining permission.

Remember to obtain permission to use copyrighted works and always credit sources that you use within the fair use guidelines. It is also a good idea to let people know when you are linking to their materials, out of common courtesy. For more information, test your knowledge with this fair use scenario in the Thought and Reflection box.

Thought and Reflection

A trainer has been told by students that it is difficult to obtain reserve materials because of the large number of students enrolled. As an alternative, he scans several journal articles onto the campus network and instructs the students on how to access them so that they may complete the class assignments. Is this fair use?

A problem with making text available on any network is that it can be accessible by readers far beyond the intended audience of students registered in the class. Thus, restrictions on access through passwords or other systems can enable the trainer to argue that the purpose is solely to benefit the students and not to provide access for others. By limiting the range of users, the trainer can minimize or eliminate the possibility that someone will retrieve the work from the network instead of purchasing a copy. One concern is that this cannot be controlled, because students could download, transmit, and share with others with little cost or effort.

Copyright Notice, (1995), The Trustees of California State University. Used by permission. http://www.cetus.org

Student Support Issues

Course developers, instructors, and trainers must have support to be successful in program planning and delivery. This is also the case for the learners at a distance. Learners must have support for academic advising/counseling, library services, training on equipment and software, financial aid, testing and assessment procedures, access to instructional resources, and guidelines on equipment requirements (Simonson & Bauck, 2003). In examining the student support services that are needed, it is necessary to examine two types of learners that we serve: primary distance-source learners and secondary distance-source learners.

Primary distance-source learners are those who are not actually tied to a campus or learning facility. They are learners with families and homes; they are motivated by a need to have access to a continuing education or formal program for personal advancement. Usually, distance learning provides a solution for them because they cannot come to the primary learning facility.

These learners have many responsibilities that they must manage on a daily basis; they are time and place bound. They are very focused on the practical aspects of learning and bring rich life experiences to the educational environment. It is usually a significant cost to continue their education and training and they will bear this cost themselves. The lure of on-site instruction is not strong; the choice to participate in the learning is based more on accessibility than content or reputation of the institution offering the instruction. Finally, they prefer not to travel unless absolutely necessary.

Secondary distance-source learners are those who are actually on campus and are seeking control of their learning environment. They are attracted to distance learning because of its convenience. They can choose when and where they want to do their course work rather than having to be in a classroom at a specific time.

Secondary distance-source learners are characterized as a generation of technology users; they have been exposed to technology since the early years of primary school. These learners seek greater control over learning experiences and typically are in close proximity to the organization sponsoring the instruction; flexibility in scheduling is a primary motivating factor.

These learners see technology as a fundamental tool to their education and training. In their opinion, no one person can have all the answers and online services are expected, and in fact, demanded. These learners will not assume that what is offered in class is the final answer; they will expect supplemental resources. They expand the educational opportunities available to them. Because of their comfort with technology, they will create networks of trainers and learners; this extends instructional strategies.

Depending on the needs of primary and secondary distance-source learners, support services may vary. Support services are organizationally sponsored functions or activities carried out for, with, or on the behalf of the learner to assist, support, and/or extend his or her educational experiences (Jackson, 2000). The same level of services is expected for both primary and secondary-source learners and on-site learners.

The Western Cooperative for Educational Telecommunications includes a comprehensive guide of necessary student support services: information for prospective students, admissions, financial aid, registration, orientation services, academic advising, technical support, career services, library service, services for students with disabilities, personal counseling, instructional support and tutoring, bookstore, and services to promote a sense of community (see Internet Connections below). This link also includes a section entitled "Two Comprehensive Web-Based Student Services Systems" that highlights ex-

Figure 3. Principles of good practices for student support services

Principles of Good Practice Related to Student Support
• Provide adequate access to the range of services appropriate to support the distance education program (registration, financial matters, delivery of course materials, etc.). • Provide an adequate means for resolving learner conflict. • Provide complete information to all learners about distance education programs, requirements, and services. • Ensure that learners have the necessary competencies and technology to take full advantage of the education program.

amples of institutions that are "customer centered" in their online service. These programs exemplify best practices (Figure 3).

There are several things you should do to take action in your organization for student support services: educate yourself on what is available; identify and communicate support needs; familiarize yourself with resource people; create linkages and partnerships; and network through committees and task forces. Student services will vary with your institution.

 **Internet Connections**

http://www.wcet.info/projects/laap/guidelines/

This site was developed by the Western Cooperative for Educational Telecommunications (WCET; aka Western Interstate Commission for Higher Education [WICHE]) and provides a comprehensive view about online student support services. Good practice recommendations and examples of institutions that are customer centered are included.

http://salc.wsu.edu/Advising/newman/default.asp

This link provides an example of advising services available at Washington State University.

http://www.walden.edu/stud-srvcs/academic-counseling

At this site, you will see an example of mentoring services available at Walden University.

http://www.crk.umn.edu/people/admissions/virtual.htm

Here is an example of counseling services at the University of Minnesota-Crookston (counseling).

Conclusion

This chapter provided an overview of key components for course and program management. Policies surrounding academic issues, faculty support and training, copyright and fair use guidelines, and student support services were emphasized. Take a few moments to think about your own industry or academic setting and create a management checklist to help guide you for course and program management in the application exercise. In the next chapter, the focus will be on course and program evaluation based upon principles of best practice.

 Application Exercise

A distance education professional must understand the basics of administrative issues. For this assignment, you will (1) create a list of the support needs of the faculty and learners (library resources, telecommunications infrastructure, software requirements, staff time, services, etc.), and (2) create a list of copyright restrictions/concerns/justifications for use of your materials and resources. See the example provided below.

Example of Checklist for Managing

Support Needs

Administrative:

- Web server administrator/technical staff: to maintain server; troubleshoot and fix technical problems with the site
- Instructional design/graphics experts
- Distance learning technology experts
- Faculty/Specialist to develop content and assessments
- Staff/Specialist to develop course evaluation instrument

- Staff to secure copyright permissions
- Staff to check and update links and information
- Staff/Specialist to access, print, read, and analyze evaluation form submissions from participants
- Staff/Specialist to make changes to course according to feedback received
- Staff/Specialist to change case studies and videos periodically

Learner:

- How to access site/URL address
- Hardware/software to access course components
- Provide text/audio/video variety
- Provide related information links
- Provide opportunity to assess knowledge
- Provide opportunity to apply information to their own life
- Provide opportunity to ask questions

Marketing:

- Provide index link from various search engines
- Provide index link from master Web sites that list online courses
- Provide index links from related topic areas

Copyright Concerns

Secure Permission to use:

- Textbook photographs
- Case studies
- Videos

Properly Credit:

- Content information – Internet resources, newsletters, agency bulletins, and so forth.

References

Bates, A.W. (1997). Restructuring the university for technological change. Retrieved from *http://bates.cstudies.ubc.ca/carnegie/carnegie.html*

Bielefield, A., & Cheeseman, L. (1997). *Technology and copyright law: A guidebook for the library, research, and teaching professions.* New York: Neal-Schuman.

Bruwelheide, J.H. (1995). *The copyright primer for librarians and educators.* Chicago: American Library Association.

Chute, A., Thompson, M., & Hancock, B. (1999). *The McGraw-Hill handbook of distance learning.* New York: McGraw-Hill.

Dooley, K.E., & Murphy, T.H. (2000). College of Agriculture faculty perceptions of electronic technologies in teaching. *Journal of Agricultural Education, 42*(2), 1-10.

Duning, B., Van Kekerix, M., & Zaborowski, L. (1993). *Reaching learners through telecommunications.* San Francisco: Jossey-Bass.

Gasaway, L.N. (2001). Balancing copyright concerns: The TEACH Act of 2001. *Educause Review, November/December,* 82-83.

Howard, C., Schenk, K., & Discenza, R. (2004). *Distance learning and university effectiveness: Changing educational paradigms for online learning.* Hershey, PA: Idea Group.

Jackson, K. (2000). *Determining student support services for distance learners in American higher education.* Unpublished doctoral dissertation, Texas A&M University, College Station.

Jones, E.T., Lindner, J.R., Murphy, T.H., & Dooley, K.E. (2002). Faculty philosophical position towards distance education: Competency, value, and education technology support. *Online Journal of Distance Learning Administration, 5*(1). Retrieved from *www.westga.edu/~distance/jmain11.html*

Kanuka, H., Collett, D., & Caswell, C. (2002). University instructor perceptions of the use of asynchronous text-based discussion in distance courses. *American Journal of Distance Education, 16*(3), 151-167.

King, J.W., Nugent, G.C., Russell, E.B., Eich, Jr., & Lacy, D.D. (2000). Policy frameworks for distance education: Implications for decision makers. *Online Journal of Distance Learning Administration, 3*(2), 1–5. Retrieved , from *www.westga.edu/~distance/jmain11.html*

Miller, G., & Pilcher, C.L. (2001). Levels of cognition researched in agricultural distance education courses in comparison to on-campus courses and to faculty perceptions concerning an appropriate level. *Journal of Agricultural Education, 42*(1), 20-27.

Moore, M.G. (1997). Quality in distance education: Four cases. *The American Journal of Distance Education 11*(3), 1-7.

O'Quinn, L., & Corry, M. (2002). Factors that deter faculty from participation in distance education. *Online Journal of Distance Learning Administration, 5*(4). Retrieved from *www.westga.edu/~distance/jmain11.html*

Perreault, H., Waldman, L., Alexander, M., et al. (2002). Overcoming barriers to successful delivery of distance-learning course. *Journal of Education for Business, 77*(6), 313-318.

Rockwell, S.K., Schauer, J., Fritz, S.M., & Marx, D.B. (1999). Incentives and obstacles influencing higher education faculty and administrators to teach via distance. *Online Journal of Distance Learning Administration, 2*(4). Retrieved from *www.westga.edu/~distance/rockwell24.html*

Simonson, M., & Bauck, T. (2003). Distance education policy issues: Statewide perspectives. In M. Moore, & W.G. Anderson (Eds.), *Handbook of distance education* (pp. 417-424). Mahwah, NJ: Lawrence Erlbaum.

Simonson, M., Smaldino, S., Albright, M., & Zavcek, S. (2003). *Teaching and learning at a distance: Foundations of distance education.* Upper Saddle River, NJ: Merrill Prentice Hall.

Strong, W.S. (1999). *The copyright book: A practical guide.* Cambridge, MA: MIT Press.

Stufflebeam, D.L. (1974). Alternative approaches to educational evaluation. In *Evaluation in education: Current applications.* Berkeley, CA: McCutchan.

Chapter XIII

Evaluating Distance Education Programs Using Best Practices

with
Kathleen Kelsey, Oklahoma State University, USA

 Making Connections

In the previous chapter, we explored a variety of administrative issues relevant to distance education. One important component is to determine if the program being offered at a distance is successful. Systematic evaluation allows the program planners and administrators to make this determination. The process of determining the merit, worth, or value of something, or the product of that process is evaluation. Terms used to refer to this process include "appraise," "analyze," "assess," "critique," "examine," "grade," "inspect," "judge," "rate," "rank," "review," "study," and "test" (Scriven, 1991, p. 139). What are indicators of quality or best practice? What tools or strategies can you use for program evaluation?

Introduction

In section III, Systematic Instructional Design, we offered suggestions for measuring learning outcomes based on the instructional objectives, in other words, learner assessment. Here, we differentiate learner assessment from program evaluation. As discussed in the previous chapter, distance education courses and programs should undergo multilevel evaluation: functional, managerial, and instructional. This chapter will operationalize models for gathering evaluation data for distance education courses and programs.

Informal evaluation is an everyday affair in our lives, both personal and professional. We make judgments of merit, worth, and value throughout our waking hours. Our ability to evaluate our environment is necessary for survival and this activity has been hardwired into our lower brain stem for millions of years.

Evaluation as a systematic process applied to educational programs became a formalized endeavor on a national basis in 1965 when President John F. Kennedy signed into law the Elementary and Secondary Education Act (ESEA) mandating a formal evaluation for all federally funded educational programs. Title 1 (later Chapter 1) of the ESEA mandated educators to evaluate their programs and was the "largest single component of the bill ... destined to be the most costly federal education program in American history" (Fitzpatrick, Sanders, & Worthen, 2003, p. 32). Since that time, hundreds of models for evaluating educational programs have been developed to address the need for accountability for government and privately funded programs.

Professional evaluators founded an organization called the American Evaluation Association and continuously improve their trade through professional development activities. Evaluators adhere to a set of guiding principles developed to support quality evaluation work and ethical behavior and apply uniform and systematic standards to conducting an evaluation.

The Program Evaluation Standards include the concepts of *utility* (will the evaluation serve the information needs of intended users?), *feasibility* (was the evaluation conducted in a realistic, prudent, diplomatic, and frugal manner?), *propriety* (was the evaluation conducted in a legal and ethical fashion, and were human subjects protected?), and *accuracy* (is the evaluation accurate and the best representation of the merit and worth of the program?).

 **Internet Connections**

www.eval.org

This link will take you to the official Web site for the American Evaluation Association.

www.eval.org/EvaluationDocuments/aeaprin6.html

To learn more about the Guiding Principles for Evaluators, check out this site.

www.wmich.edu/evalctr/jc/

To review the Program Evaluation Standards, this link is an excellent resource.

Evaluating distance education programs formally and in a systematic fashion is critical for program improvement (formative) and documenting successes (summative). Without evaluation, we have no way of knowing empirically if our programs are having their desired impact. In this chapter, we will first discuss best practices for distance education and then outline a method for determining if your program is in alignment with best practices.

Best Practices for Distance Education

The most frequently cited reference for best practices comes from a joint publication of the Western Cooperative for Educational Telecommunications (WCET), otherwise known as Western Interstate Commission for Higher Education (WICHE). WICHE was established by the Western Regional Education Compact in the 1950s for the purpose of facilitating resource sharing among its member states. The Commission has published a number of documents including *Good Practices in Distance Education* (WCET, 1997). WICHE's best practices for distance education fall under three principles: (1) curriculum and instruction, (2) institutional context and commitment, and (3) evaluation and assessment.

Adding to WICHE's three best practices domains, Law, Hawkes, and Murphy (2002) have outlined general questions and concerns under five headings that should be addressed when evaluating a distance education program. The five constructs along with possible questions to consider are outlined below.

Consistency of Program With Institutional Mission

Is the distance education program consistent with the role and mission of the institution? Budgets should reflect institutional commitment to the distance education degree program as well as technological infrastructure such as bandwidth and server capacity. Are the reasons why learners enroll in the program consistent with the goal and mission of the program? Evaluators should look for evidence that offering distance education is in line with the institutional mission and is well supported with adequate budgets and support staff.

Provisions for Program Oversight and Accountability

Is there an administrative office that is responsible for the distance education program? Is there infrastructure in place to ensure the quality of the program? Or are individual instructors struggling alone to deliver courses with no centralized support? Is there a unique curriculum for the distance education program that originated at the degree-granting institution or training center? Do learners have access to all aspects of the program required for success? Is there satisfactory coordination of different aspects of the program? Academic and technical oversight and accountability should be obvious to evaluators.

Provisions for Instructor Support

Have instructor responsibilities and compensation been rebundled to respond to the demands of teaching or training at a distance? Or are instructors simply required to add distance teaching responsibilities to their existing workload? Evaluators should look for evidence of restructuring of instructional positions to reflect the unique needs of teaching or training at a distance.

Provisions for Learner Support

Does the institution provide support services to distance learners? For example, are workable arrangements in place for using library facilities in different locations? Are "hassle-free" administrative (e.g., admission and registration) procedures in place? Are adequate provisions and procedures for financial assistance made as needed? Are schedules such that an adequate amount of time is spent interacting with learners? Learner support goes far beyond the relationship between instructors and learners to include "advising, placement, enrollment, financial aid, tutoring, and technology assistance" (Law, Hawkes, & Murphy, 2002, p. 87). Evaluators should examine Web sites and other media to ensure that learners have access to all the required contexts for learning online.

 **Internet Connection**
http://www.wiche.edu

This Web site takes you to the Western Interstate Commission for Higher Education's Good Practices in Distance Education.

Implementation of Evaluation and Assessment Measures

Is the program achieving its learning objectives? Are the participants learning? Is the curriculum/content rigorous? Are learner assessments authentic? Are learners and instructors satisfied with the program? Are the learner completers or employers of learner completers recommending the program and its courses to others? Do employers of learners who take specific courses reward them after completion of the courses? Are learners using the preparation, knowledge, and skills gained from distance courses in their field of endeavor? Evaluation is a critical component of an excellent distance education program and should not be an afterthought, but rather should be incorporated into the planning phase of the program from inception.

The Western Association of Schools and Colleges (WASC) (2001) provides a self-assessment framework for determining quality in distance education

Table 1. Components of good practice

Component of Good Practice	WASC Interpretation of Good Practice
Institutional context and commitment	"Electronically offered program both support and extend the roles of educational institutions. Increasingly they are integral to academic organization, with growing implications for institutional infrastructure." (p. 1)
Curriculum and instruction	"Methods change, but standards of quality endure. The important issues are not technical but curriculum-driven and pedagogical. Decisions about such matters are made by qualified professionals and focus on learning outcomes for an increasingly diverse student population." (p. 4)
Faculty support	"Faculty roles are becoming increasingly diverse and reorganized. For example, the same person may not perform both the tasks of course development and direct instruction to students. Regardless of who performs which of these tasks, important issues are involved." (p. 8)
Student support	"Colleges and universities have learned that the twenty-first century student is different, both demographically and geographically, from students of previous generations. These differences affect everything from admissions policy to library services. Reaching these students, and serving them appropriately, are major challenges to today's institutions." (p. 9)
Evaluation and assessment	"Both the assessment of student achievement and evaluation of the overall program take on added importance as new techniques evolve. For example, in asynchronous programs the element of seat time is essentially removed from the equation. For these reasons, the institution conducts sustained, evidence-based and participatory inquiry as to whether distance learning programs are achieving objectives. The results of such inquiry are used to guide curriculum design and delivery, pedagogy, and educational processes, and may affect future policy and budgets and perhaps have implications for the institution's roles and mission." (p. 12)
Retrieved from www.wascweb.org/senior/distance_ed_resource.htm	

programs. WASC divides good practices into five components. Table 1 details the five components of good practice and discusses the WASC interpretation of each component. Please note the similarity with the five categories outlined above by Law, Hawkes, and Murphy (2002).

A faculty consortium from Pennsylvania State, Lincoln, and Cheyney Universities developed guiding principles for distance education. They are consistent with principles of good practice and include (1) learning goals and content presentation, (2) interactions, (3) assessment and measurement, (4) instructional media and tools, and (5) learner support systems and services.

Rocha-Trindade, Carmo, and Bidarra (2000) provide a comprehensive review of best practices for open and distance learning and suggest that program planners assure quality by examining learning materials "with respect to both their scientific content and pedagogic strategies," in addition to evaluating "student support mechanisms, communications, and organization and logistics" (p. 15).

 Internet Connections

http://www.electroniccampus.org/student/srecinfo/publications/Principles_2000.pdf

Go to this site for a review of the SREB's Principles of Good Practice.

http://www.educause.edu/ir/library/html/cem9915.html

Review principles and practices for the design and development of distance education programs developed by Pennsylvania State, Lincoln, and Cheyney Universities.

http://www.irrodl.org/content/v1.1/carmo_et_al.html

Review current development and best practice in open and distance learning.

Using Best Practices as a Standard for Program Evaluation

Given the variety of distance education programs available today, from offering one course to an entire degree program, program planners should not adopt a standardized instrument to use for evaluating their program, no matter how tempting it may seem to be at the time. For example, the authors know of a case in which an instrument developed, validated, and used successfully in one state to evaluate programs in that state was adopted lock, stock, and barrel without modification (the cover only was changed from blue to tangerine) and used in another state where both program conditions and the process used to collect evaluation data were different. The result was that instructors received poor evaluations of their programs; program evaluation became a dirty word, and now, many years later, there is still resistance among the veteran instructors in that state to formal program evaluation. Rather, program planners should consider the best practices literature as a guide to developing an evaluation plan that addresses the unique context and setting for their particular distance education program (Law, Hawkes, Murphy, 2002).

Best practices for distance education are provided as a guide and can be used as a gold standard for evaluating programs. The underlying questions to ask when designing an evaluation should be, "Is this particular standard relevant to my program? If so, to what degree of quality should it be operationalized?"

If the standard is not present in your program, then ask why not? Not every program will have every best practice, and for good reason. Not all practices are appropriate for all programs. Only the program planners and decision makers know the reasons behind including or excluding specific recommendations from the literature on best practices. The evaluator's role is to make a salient argument for such inclusion or exclusion given the uniqueness of each program.

Evaluation Models

There are hundreds of books written about program evaluation and hundreds of thousands of journal articles documenting evaluation practice. A simple Internet search using the term "program evaluation" will yield over three million hits. Which model is right for your program? This section will outline Stufflebeam's (1973) Context, Input, Process, and Product (CIPP) evaluation model, Chen's (1990) theory-driven evaluation model, and the logic model process.

The CIPP model developed by Stufflebeam (1973) includes four phases of evaluation. Phase one is *context* centered and addresses the questions of where is your program now, what are your program's needs, and where do you want your program to be? Phase two is *input* centered and asks the questions, "How will you get where you want to be?" and "What resources are required to drive your program?" Phase three concerns the *process* (called "ways and means" in other evaluation models) and asks how are you going to achieve your program goals? Phase four is *product* focused (called "outcomes" in other evaluation models) and asks if your program has achieved its goals and what are the outcomes (Table 2).

The CIPP model was developed during the early years of the program evaluation discipline and has been refined several times by various authors to be more user friendly. Theory-driven evaluation was developed by Chen (1990) over the next decade and indirectly explains the context as the program's *implementation environment*, the input as the program's *treatment*, the process as the program's *intervening mechanisms*, and the product as the *outcomes* of the program. Chen asks evaluators and program stakeholders to reflect on the cause-and-effect mechanisms for each program. What are the causal elements that drive behavior change (learning in the case of distance education) and what are the effects, or outcomes, of the program's treatments (teaching/training, assignments, learner–instructor interactions)? Figure 1 depicts Chen's model.

Logic modeling has been discussed by a number of authors and is once again a refinement of Stufflebeam's (1973) CIPP model and Chen's (1990) work. A logic model is defined as a "systematic and visual way to present and share your understanding of the relationships among the resources you have to operate your program, the activities you plan to do, and the changes or results you hope to achieve" (W.K. Kellogg Foundation, 2001, p. 1). Figure 2 depicts a flow of activities required to develop a logic model for a program.

Table 2. Stufflebeam's four phases of evaluation

Phase	Components of the Phase
Phase one	Context centered and addresses the questions of where is your program now, what are your program's needs, and where do you want your program to be?
Phase two	Input centered and asks the questions, "How will you get where you want to be?" and "What resources are required to drive your program?"
Phase three	Process centered and asks how are you going to achieve your program goals?
Phase four	Product centered and asks if your program has achieved its goals and what are the outcomes.

Figure 1. Chen's model for theory-driven evaluation (1990, p. 50)

![] **Internet Connections**

http://www.wkkf.org/Pubs/Tools/Evaluation/Pub3669.pdf

Go to this site to obtain a copy of W.K. Kellogg Foundation's (2001) Logic Model Development Guide.

http://citnews.unl.edu/TOP/english/index.html

You may access the TOP evaluation model at this site. The Targeting Outcomes of Programs (TOP) Model was developed by leading thinkers Claude Bennett and Kay Rockwell in education and evaluation and has been used for the past 20 years.

Logic modeling was intended as a group activity for program planners and decision makers to conceptualize and verbalize their program's context, inputs (resources/inputs), processes (activities), and products (outcomes and impact). Once all stakeholders agree on the core program elements (CIPP) the evaluation can begin to determine if goals have been achieved. Logic modeling is a very functional and important process for program stakeholders to undertake and complete before a summative evaluation is commissioned. Logic modeling will illuminate people's construction of the program goals, values, processes, and expected outcomes.

Figure 2. The basic logic model

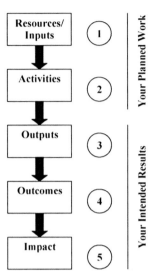

A Simple Plan of Action for Evaluating Your Distance Education Program

Regardless of the evaluation model chosen for evaluating the program, stake-holders should remain at the center of all processes. Stakeholders are those people who care about your program, including learners (beneficiaries), instructors, program planners, decision makers, technicians, and funders (agents). Another category of stakeholder that must not be forgotten are those who do not benefit from the program or are harmed by the program such as learners who are not admitted to the program (victims) (Guba & Lincoln, 1989). Including all stakeholders (beneficiaries, agents, and victims) in the design and implementation of a program evaluation is the first principle of evaluation practice (Bryk, 1983).

Boulmetis and Dutwin (2000, p. 70) suggest a seven-step approach for conducting evaluation: (1) determine evaluation questions, (2) develop the evaluation design, (3) collect data, (4) analyze data, (5) draw conclusions from data, (6) make decisions on a program's efficiency, effectiveness, and impact, and (7) report to stakeholders. While this is a sound design, an evaluation that includes stakeholders in all steps will increase the likelihood that the results will be eagerly read and acted upon by program planners and decision makers.

The following evaluation design is proposed to be stakeholder centered and emphasizes use of evaluation findings to improve distance education programs:

1. Realize that evaluation is a *group process* driven by people who are passionate about creating the best distance education program possible. The role of the evaluator is to facilitate the process, not to direct or control it. The evaluation belongs to the program planners, decision makers, instructors, learners, administration, and other stakeholders, not the evaluator.

2. Identify and make a list of stakeholders (beneficiaries, agents, and victims).

3. Gather interested stakeholders to determine the focus of the evaluation. Meaningful evaluation is a group process. What do stakeholders want to learn about their program? What elements of the program need improvement? The brainstorming session can begin with the question, "I want to know _____ about our distance education program" (Patton, 1997).

4. Develop evaluation questions that are important for improving the program and that can be answered.

5. With the help of a social science methodologist, develop a plan of action for collecting credible evidence (data) to answer your evaluation questions (see Tables 3 and 4 for additional information on data collection methods).

6. Share all processes, procedures, and documents with stakeholders and gather feedback about the proposed methods before collecting any data. The evaluator needs complete agreement from stakeholders before collecting any data. Data that are not desired will be data that collect dust on the shelf.

7. Once stakeholders agree on the right questions to ask and the right way (methods) to ask questions, data collection may begin.

8. Involve as many interested stakeholders as possible in collecting the data. People who care about the evaluation will use the results for program improvement (Patton, 1997).

9. Once the data is collected, assemble a team to analyze the data. This step may require the assistance of a social science methodologist for statistical procedures or for synthesizing volumes of interviews, notes, and observations, or both.

10. Share all findings early and often. Keep the flow of information to stakeholders continuous, even if your report is only in draft form. Do not let the fires die down on the evaluation process as interest will wane if there are time lags between communications. In fact, circulate drafts of the findings for correction and clarification among participants. Include a cover letter that stresses the value of getting the facts right by asking participants to read the analysis and provide feedback. The more stakeholders are involved in the entire process, the more valuable the findings will become to the users of the information.

11. Once all stakeholders have reviewed the draft report and given their approval of the content, the evaluator should develop a plan of action for disseminating the final report. At this stage in the process, most stakeholders will have read several drafts of the report, will have discussed the findings, and will have begun the process of using the evaluation findings for program improvement so that a report is merely a formality for archiving the evaluation. Keep in mind that a written report may be the least used document in your organization. Other approaches include having a meeting to discuss findings. Newer, more avant-garde practices for dissemination of evaluation results are performance based, such as putting on a skit, writing a poem, creating a play, or some other way of connecting stakeholders to the findings in a meaningful way. Findings that connect to stakeholders' emotions or values are findings that are remembered.

12. Follow up. A meta-evaluation is an evaluation of the evaluation. Did the evaluation make a difference in practice? If yes, communicate your success to stakeholders. They need to understand that their time and effort were put to good use. This will make the next evaluation a breeze! If no, figure out why and evaluate again. Remember, evaluation that is worthwhile is a continual process.

Evaluation Methods for Collecting Credible Evidence

The purpose of any method is to collect *credible evidence* to document program activities. Information is used to make decisions for program improvement. When seeking methods for the evaluation, consider what information is needed to make decisions about the program versus the cost and ease of collecting that information.

Table 3. Major methods used for collecting evaluation data (McNamara, 2004)

Method	Overall Purpose	Advantages	Challenges
Questionnaires, surveys, checklists	When need to quickly and/or easily get large amount of information from people in a nonthreatening way	-Can be complete anonymously -Inexpensive to administer -Easy to compare and analyze -Administer to many people -Can get large amount of data -Many sample questionnaires already exist	-Might not get careful feedback -Wording can bias client's responses -Are impersonal -In surveys, may need sampling expert -Does not get full story
Interviews	When want to fully understand someone's impressions or experiences, or learn more about his or her answers to questionnaires	-Get full range and depth of information -Develop relationship with client -Can be flexible with client	-Can take much time -Can be hard to analyze and compare -Can be costly -Interviewer can bias client's responses
Documentation review	When want impression of how program operates without interrupting the program; is from review of applications, finances, memos, minutes, and so forth.	-Get comprehensive and historical information -Does not interrupt program or client's routine in program -Information already exists -Few biases about information	-Often takes much time -Info may be incomplete -Need to be quite clear about what you are looking for -Not flexible means to get data; data restricted to what already exists
Observation	To gather accurate information about how a program actually operates, particularly about processes	-View operations of a program as they are actually occurring -Can adapt to events as they occur	-Can be difficult to interpret seen behaviors -Can be complex to categorize observations -Can influence behaviors of program participants -Can be expensive
Focus groups	Explore a topic in depth through group discussion, for example, about reactions to an experience or suggestion, understanding common complaints, and so forth; useful in evaluation and marketing	-Quickly and reliably get common impressions -Can be efficient way to get much range and depth of information in short time -Can convey key information about programs	-Can be hard to analyze responses -Need good facilitator for safety and closure -Difficult to schedule six to eight people together
Case studies	To fully understand or depict client's experiences in a program, and conduct comprehensive examination through cross-comparison of cases	-Fully depicts client's experience in program input, process, and results -Powerful means to portray program to outsiders	-Usually quite time consuming to collect, organize, and describe -Represents depth of information, rather than breadth

Ideally, a variety of methods should be used in combination with each other to get a complete picture. For example, using a learner survey to determine learner satisfaction with the program can be complemented with employer interviews seeking information about the quality of graduates and their preparedness for the workplace. The following table was developed by McNamara (2004) and gives a quick overview of the major methods used for collecting data (Table 3).

Another helpful resource for determining appropriate methods for collecting evaluation data was developed by Dr. Brenda Seevers of New Mexico State University (Seevers, 2004). Table 4 details the advantages and disadvantages of nine different methods for collecting data.

Table 4. Advantages and disadvantages of nine different methods for collecting evaluation data (Seevers, 2004)

Method	What it measures	Advantages	Disadvantages
Existing information	records, files, receipts, historical accounts, personnel records, reports, etc.	• readily available • minimal cost • data available on a wide variety of characteristics • can be accessed on a continuing basis • descriptive data • can provide insight into program that cannot be observed in any other way	• user may need to sort, discriminate and correlate • takes time • figures may represent estimates rather than actual accounts • does not reveal individual values, beliefs, or reasons underlying current trends
Case studies	the experiences and characteristics of selected persons in a project; generally utilized with a small number of individuals or groups	• procedures evolve as work progresses, no confining categories of classifications • allows in-depth insight into relationships and personal feelings • can be effectively used in combination with other methods, such as survey and observation • unique opportunity to study organization, group • can be tailor made to specific situations	• requires absolute accuracy • can be very subjective • time consuming, requires extensive amounts of data • focus is on a limited number of cases; cannot necessarily be generalized to larger community • not suitable as a method in isolation, best for a back-ground or as a guide to further study • several cases are needed for best analysis

Table 4. Advantages and disadvantages of nine different methods for collecting evaluation data (Seevers, 2004) (cont.)

Method	What it measures	Advantages	Disadvantages
Surveys (includes personal interviews, drop-off questionnaire & telephone interviews.)	opinions, attitudes, beliefs, behaviors, reactions, and attributes in response to specific questions	• can be inexpensive • sample can be used to provide much information about a population • can provide an opportunity for many people to be involved in the decision-making process • can be used to record behaviors as well as opinions, attitudes, beliefs and attributes • usefulness enhances if combined with other methods	• samples must be carefully selected to ensure statistical meaning. • subject to misinterpretation, depending on how questions are designed and asked • tendency for scope of data to be limited, omission of underlying behavior-al patterns • time-consuming compared with less formal methods
Mail surveys	opinions, attitudes, beliefs, behaviors, reactions, and attributes in response to specific questions	• efficient for volume of information collected • people more likely to provide frank, thoughtful, honest information, tension-free situation • gives people more time to complete the questionnaire • all respondents receive same questions in printed form	• low response rate • one or two follow-ups are usually needed for a good return • questionnaire must be simple and easy to understand • difficult to summarize open-ended questions • accurate mailing lists are required • overuse of this method may make some people reluctant to respond • privacy, confidentiality, and anonymity must be assured • can be expensive • scope is limited • results may be misleading if only respondents who are interested in topic respond

Table 4. Advantages and disadvantages of nine different methods for collecting evaluation data (Seevers, 2004) (cont.)

Method	What it measures	Advantages	Disadvantages
Telephone surveys	opinions, attitudes, beliefs, behaviors, reactions, and attributes in response to specific questions	• response rate is generally high • cost is competitive with mail survey • speedy and efficient source of data • researcher can provide clarification on unclear questions • respondents are more relaxed with a stranger by telephone than face to face • interviewer can read questions from script and take notes without concern of distracting respondents • respondents cannot read the interviewer's body language	• time consuming • telephone numbers are needed • proportion of unlisted numbers or households without phones may cause frame error • questions should still be simple and easy to understand (no more than five response categories) • interviewer's voice or identity may lead to some biasing • respondents may feel interview is an invasion of privacy • interviewer has little opportunity to "loosen up" the respondent • interviewer cannot read respondents' body language • scope of survey is limited • interviewer training may be necessary
Group administered survey	opinions, attitudes, beliefs, behaviors, reactions, and attributes in response to specific questions	• can collect a lot of data inexpensively by having everyone at a meeting or program complete the survey form • easy to clarify items which present difficulty • provides greatest sense of respondent anonymity • good method to collect baseline and attitudinal data • high response rate • can be used for quantitative and qualitative methods	• may require the cooperation of others (i.e., school administrators, etc.) • reach only those who are present • group dynamics may affect individual responses • opportunity for researcher influence

Table 4. Advantages and disadvantages of nine different methods for collecting evaluation data (Seevers, 2004) (cont.)

Method	What it measures	Advantages	Disadvantages
Personal interviews	person's responses and views	• easier to reach those who are considered unreachable (the poor, homeless, high status, mobile, etc.) • may be easier to reach specific individuals (i.e., community leaders, etc.) • higher response rate • more personalized approach • easier to ask open-ended questions, use probes and pick up on nonverbal cues • qualitative or quantitative	• may be most expensive method • slowest method of data collection and analysis • responses may be less honest and thoughtful • interviewer's presence and characteristics may bias results • interviewer should go to location of respondent • respondents who prefer anonymity may be inhibited by personal approach • may reach only a small sample
Group interviews	person's responses and views	• less expensive and faster than personal interviews • personalized approach • group members stimulate each other	• respondents who prefer anonymity may be inhibited by personal approach • input may be unbalanced because some group members dominate • group members and interviewer can bias responses • data more difficult to analyze and summarize

Conclusion

Program evaluation is a systematic process to determine the merit, worth, or value of an object, product, or process. Evaluation has evolved into a professional discipline whose members possess content-specific skills and

Table 4. Advantages and disadvantages of nine different methods for collecting evaluation data (Seevers, 2004) (cont.)

Method	What it measures	Advantages	Disadvantages
Observation	particular physical and verbal behaviors and actions	• setting is natural, flexible and unstructured • evaluator may make their identity know or remain anonymous • evaluator may actively participate or observe passively • can be combined with a variety of other data collection methods • most useful for studying a small unit such as a classroom	• evaluator has less control over the situation in a natural environment • Hawthorne effect, if group is aware that they are being observed, resulting behavior may be affected • observations cannot be generalized to entire population unless a plan for representativeness is developed • if observer chooses to be involved in the activity, they may lose objectivity • not realistic for use with large groups
Mass media and public hearings	opinions, ideas	• citizens have an opportunity to respond • teleconferencing, call in, and town meeting methods are quick methods of obtaining input	• extremes of a population tend to respond, those definitely for or against the program • use of public television and teleconferencing is limited to those who have access to public television and a phone • public hearings are time consuming, especially for the interviewers • summary and analysis of data can be difficult • response to public hearings is affected by location, distance and date

knowledge in qualitative and quantitative research methods and can even be a member of a professional organization (the American Evaluation Association). Evaluators should adhere to the guiding principles for evaluators and practice the program evaluation standards developed by leading thinkers of the profession.

Many organizations have developed best practices for distance education programs. The most commonly cited document is the WICHE guide for Good Practices in Distance Education (WCET, 1997). Best practice guides facilitate evaluating distance education programs by providing a standard for judging quality programs. Several organizations have developed checklists for evaluation purposes.

The CIPP evaluation model (Stufflebeam, 1973) is an excellent starting point for unraveling the complexities of developing and evaluating a distance education program. Chen (1990) added to the literature on evaluation models by asking evaluators to think in terms of cause-and-effect mechanisms for programs. Logic models should be used to facilitate group discussion of what are the program's cause-and-effect mechanisms, realizing that all programs are not implemented as planned, but evolve over time.

Evaluating a distance education program can occur during the program (formative) or after a program has reached an ending point (summative). Evaluation processes should include group processes and thus should include all stakeholders (agents, beneficiaries, and victims) as well as use sound research methods. Data collection should include a variety of methods such as questionnaires, checklists, observations, interviews, and focus groups. Evaluation findings should be used to improve programs and should be shared with all concerned parties.

 Application Exercise

Create an evaluation instrument to use for your lesson. It can be functional, managerial, or instructional. Decide which evaluation model you will use. Now, decide which method(s) you will use for data collection.

References

Boulmetis, J., & Dutwin, P. (2000). *The ABCs of evaluation: Timeless techniques for program and project managers.* San Francisco: Jossey-Bass.

Bryk, A.S. (Ed.). (1983). *Stakeholder-based evaluation.* San Francisco: Jossey-Bass.

Chen, H.T. (1990). *Theory-driven evaluations.* Thousand Oaks, CA: Sage.

Fitzpatrick, J.L., Sanders, J.R., & Worthen, B.R. (2003). *Program evaluation: Alternative approaches and practical guidelines* (3rd ed.). New York: Longman.

Guba, E.G., & Lincoln, Y.S. (1989). *Fourth generation evaluation.* Newbury Park, CA: Sage.

Law, J., Hawkes, L., & Murphy, C. (2002). Assessing the on-line degree program. In R.S. Anderson, J.F. Bauer, & B.W. Speck (Eds.), *Assessment strategies for the on-line class: From theory to practice* (Vol. 91, pp. 83-89). San Francisco: Jossey-Bass.

McNamara, C. (2004). Overview of methods to collect information. Retrieved January 23, 2004, from *www.mapnp.org/library/evaluatn/fnl_eval.htm#anchor1585345*

Patton, M.Q. (1997). *Utilization-focused evaluation: The new century text* (3rd ed.). London: Sage.

Rocha-Trindade, A., Carmo, H., & Bidarra, J. (2000, June). Current developments and best practice in open and distance learning. *International Review of Research in Open and Distance Learning.* Retrieved January 14, 2004, from *www.irrodl.org/content/v1.1/carmo_et_al.html*

Scriven, M. (1991). *Evaluation thesaurus* (4th ed.). Thousand Oaks, CA: Sage.

Seevers, B. (2004). Advantages and disadvantages of nine different methods for collecting evaluation data. (Available from the author, New Mexico State University, Gerald Thomas Hall Room 111, P.O. Box 30003 MSC 3501, Las Cruces, NM 88003)

Southern Regional Educational Board's (SREB's) Electronic Campus. (2003). Principles of good practice. Retrieved January 26, 2004, from *www.electroniccampus.org/student/srecinfo/publications/Principles_2000.pdf*

Stufflebeam, D.L. (1973). *Toward a science of educational evaluation.* Englewood Cliffs, NJ: Educational Technology Publications.

Western Association of Schools and Colleges (WASC). (2001). *Good practices for electronically offered degree and certificate programs.* Retrieved January 26, 2004, from *www.wascweb.org/senior/distance_ed_resource.htm*

Western Cooperative for Educational Telecommunications (WCET). (1997). *Balancing quality and access: Principles of good practice for electronically offered academic degree and certificate programs.* Pub. #2A299. Retrieved January 9, 2005, from *http://www.wcet.info/projects/balancing/principles.asp*

W.K. Kellogg Foundation. (2001). Logic model development guide. Retrieved February 19, 2004, from *www.wkkf.org/Pubs/Tools/Evaluation/Pub3669.pdf*

Section VI

Future Directions

Although we cannot predict the future, we believe that distance education will be shaped by educators who want to efficiently and effectively reach distance learners across the globe. This section will focus on the educational and technological trends that will impact the future. Some visions of the future are included as well as ways to stay current in the field.

Chapter XIV

Future Trends

with
Chehrazade Aboukinane, Texas A&M University, USA

 Making Connections

In this book, so far, we have covered five major sections: Foundations of Instruction and Learning at a Distance, Adult Learning Theory, Systematic Instructional Design, Technology Knowledge and Skills, and Administrative Issues. We based the chapters in these sections on seminal and current research, as well as practical applications in higher education and human resource development. The final chapter of this book explores future trends and directions in the field of distance education. We know that technology continues to increase in power while decreasing in cost, so it is clear that the technological capabilities will change. But will adult learning theory and instructional design principles also change? We stressed that these fundamental principles will remain constant and serve as the foundation for effective instruction and learning, but who knows what the future holds? What major changes will occur in the field of distance education 10, 20, or even 50 years from now?

Introduction

When we paused to reflect on possible futures for the field of distance education, we chose to begin with what other professionals predicted in their writings. In an article created for the *Quarterly Review of Distance Education*, Wilson (2001) stated that technologies are still reflections of us:

> Through technologies and new ideas, we are always in the process of re-inventing ourselves. Technologies serve as mirrors of our values and aspirations, as well as our weaknesses and intractable problems. This truth about technologies underscores the importance of subjecting our plans to continuing scrutiny. Whenever possible, we want our technologies to reflect our best selves and our highest ambitions (p. 12).

With this is mind, we explored both the educational changes and the technological advances that will impact instructing and learning at a distance in the future.

Educational Trends

Education is changing as a result of distance learning applications and practices. Howell, Williams, and Lindsay (2003) conducted a literature review of distance education journals to compile a list of 32 trends that will impact distance learning. This list was specific to higher education but can be applied to other settings. The trends were clustered into student/enrollments trends, faculty trends, academic trends, technology trends, economic trends, and distance learning trends. Can you make a list, based on your own experience, of what some of the 32 trends are likely to be? Incidentally, you might identify some trends not mentioned that if you reviewed such a list of trends in five years or so, you could say, "I told you so!" To read the complete article to find out what others have predicted, check out the Internet Connection on the next page.

Howell, Williams, and Lindsay demonstrate the interrelationships of future trends that will impact distance education. Good (1999) also emphasized the

Internet Connection

http://www.westga.edu/~distance/ojdla/fall63/howell63.html

This site is linked to the article "Thirty-Two Trends Affecting
Distance Education: An Informed Foundation for Strategic Planning"
by Scott Howell, Peter Williams, and Nathan Lindsay. It can be found
in the *Online Journal of Distance Learning Administration*.

influence of additional factors on educational trends. "Education is not an island; it is affected not only by what is happening in the field but also by what is happening in the rest of society—population changes, technological advances, economic ups and downs, political shifts and social transformations" (Good, 1999, p. 5). In his article, "Future Trends Affecting Education," he predicts more competition among educational providers and a need for more accountability at all levels. Learners will become increasingly more diverse. The demand for technically skilled workers will increase, which in turn, will impact the need for distance education (Good, 1999). The Internet Connection below provides the link to this article.

 Internet Connection

http://www.ecs.org/clearinghouse/13/27/1327.htm

The Education Commission of the States developed *Future Trends
Affecting Education* penned by Dixie Griffin Good, an education
consultant.

Other views focus on how education is changing due to technological advances. It is widely known that the development of personal computers and the Internet has improved our ability to share intellectual and social resources across vast distances and improves our access to lifelong learning. Castro (2001) predicts an increase in Web-delivered courses, changing instructional roles, and centralization of curriculum development. Another educational futurist, Thornburg (1998), worries about the lack of technologically competent workers and preparing for jobs that do not currently exist. He claims that as "technologies become commonplace with all students, the tools for lifelong learning will be in place" (p. 6). The technology tools we use for distance education will reduce in cost and make learning more engaging and knowledge more accessible (Snyder, 2004).

According to Wilson (2001), technology and ideas will continue to coevolve:

> Historians of technology tell us that a technology, often based on the best thinking available, in turn stimulates new thinking and new possibilities. This is certainly true of the Web and networked information systems. A huge spike of promising ideas, models, and R&D efforts has accompanied the new technology. When these new efforts are seen as artifacts themselves, we see how one technology prompts the development of another, and how the cycle repeats itself through new iterations of technology, design, theorizing, and practice. Thus we can be sure that, as technology continues its onward march, new models and ideas will surely follow—and in some cases, precede the technology itself (p. 12).

Technological Advances for Distance Learning

Draves in 2000 dedicated a chapter in *Teaching Online* to "How the Internet Will Change How We Learn." He believed that online classes will have from 100 to 1,000 participants while maintaining more interaction among instructors and learners. The learners and instructors will not be restricted by location and will form a virtual community from around the world. Although this might be likely in the future, many universities limit enrollment in their online classes to approximately 20 learners, including the University of Phoenix online which instructs close to 100,000 distance learners.

Chute, Thompson, and Hancock (1999) discuss the future of networked learning environments. These authors foreshadow the extended use of bookmarking for virtual journeys with intelligent "agents" that "remember" learners' interests and requirements (p. 216). These environments will include three-dimensional (3-D) presentations, artificial intelligence, and virtual reality. Virtual reality (VR) is the use of computers to simulate a real or imagined environment that appears as a 3-D space. VR allows you to explore and manipulate controls to experience the 3-D space completely (Thomson Course Technology, 2003). This can have implications for learning complex motor skills (psychomotor domain of learning) in distance training.

Negroponte states that bits are the "DNA of information" and are replacing atoms as the "basic commodity of human interaction" (1995, p. 6). He also predicted that computers would soon "be freed from the confines of keyboards and screens" and instead would be "objects that we talk to, drive with, touch or even wear" (p. 6). According to Thomson Course Technology, new wireless products will allow users to strap sensors on their fingers or hands and then make hand motions to type or control their computers.

Other technologies that may impact distance learning is speech recognition, wireless technology, and mobile Internet access. Speech recognition allows a computer to translate spoken statements into computer instructions. The market for voice recognition software has skyrocketed to about $22 billion, up from $356 million in 1997. Microsoft Office XP includes speech recognition software (Thomson Course Technology, 2003).

Learners are looking for new ways to access the Internet for communication. Wireless technology is in high demand. Advances in digital wireless technology have focused on using mobile telephones and handheld computers and devices. The next logical step is cordless computers. Wrist-top computers will integrate telephone, e-mail, pager messages, and appointments. Palmtops are expected to be the top-selling wireless devices with an average annual growth of 28% through 2004. Other advances are expected in use of global positioning systems, electronic paper and ink, and multiple-function telephones (Thomson Course Technology, 2003).

Online learners will also take advantage of a variety of mobile Internet services. For example, a mobile Internet service called "i mode," created by DoCoMo, Inc., a Japanese mobile Internet company, is predicted to lead broadband services worldwide (Thomson Course Technology, 2003).

Other technology advances that will facilitate distant learning include portable devices. Manufacturers are constantly developing new handheld devices, such as microdisplays and headsets, that will pervade the world of distance education. Microdisplays are the size of postage stamps and will allow distance learners to view lessons, Web pages, e-mail, and all course-related materials. Microdisplays can appear as large as a regular monitor when magnified. Even sunglasses can be used as learning tools. IBM has recently developed "learning glasses" that allow translating Japanese signs into English.

It is evident that technological advances will continue to play a part in the development and integration of distance education learning environments. Now, let us review some learner and educator perspectives on future trends.

Visions of the Future of Distance Education From Varying Perspectives

When we paused to reflect on future trends and directions of distance education, we felt that our view would be somewhat limited by our own experiences and biases. So we thought you would gain more insight by reading responses from varying perspectives of people in different settings. We surveyed graduate students, professors from around the United States, professionals working in extension, training, and development, and colleagues from around the globe to determine their vision for distance education in the future.

We explained that we were writing a book on designing and managing distance education programs and courses and wanted to solicit their input on the future of distance education. Specifically we asked them to write their thoughts about future trends with respect to any of the following items: the role of adult learning, self-directedness, instructional design, interactions, delivery strategies, multimedia, course/program management, budgeting/staffing, accountability and evaluation. Items that addressed international and cross-cultural components were particularly encouraged.

We were attempting to capture their words and thoughts about where things are going 20, 50, or even 100 years down the road. As you might guess, we captured some fascinating comments and have only included a portion—the ones we found most interesting, useful, and divergent in viewpoints. These have been organized into vignettes and representative samples will be included in the Thought and Reflection boxes to follow.

Distance Education in the Future: The Learner Viewpoint

Graduate students in the course "Advanced Methods in Distance Education" were asked via threaded discussion for their views on the future of distance education. Perspectives were varied, but several themes emerged: (1) the need for more courses/programs to meet the demand for both formal degree programs and continuing education; (2) the ability to continue to serve people who cannot participate in face-to-face delivery for a variety of reasons; (3) the ability of corporations, government, and other institutional entities to train personnel without the need for expensive travel; (4) the ability to provide global expertise and interactions with diverse populations; (5) the need for the learners to be self-directed in their approaches to learning; (6) the continued need for faculty training and learner support; (7) obvious changes in delivery strategies and technologies, yet a need to focus on best practices (pedagogy/andragogy); and (8) concerns about access to education due to financial restraints or limited bandwidth/technological capabilities.

Thought and Reflection: Student View

Distance education will become an archaic term. Ubiquitous education will replace it. As Einstein communicated, "all things are relative." When I am engaged in "distance education," I cannot tell that I am some great distance from the source of the material. It feels to me that the tools, information, people, and so forth, are a "mouse-click away."

Changing technology will always be an issue in ubiquitous education. Different institutions and different departments within institutions will continue to use varying delivery strategies. There will continue to be competition in development of the "best" delivery strategies. While this will always challenge the learner, it is a necessary challenge which yields better resources.

Ubiquitous education systems will never be developed to meet the needs of all people. But as ubiquitous educators develop the science of ubiquitous education, systems will be developed to serve the majority of potential learners.

Mike Farrow, Graduate Student, Texas A&M University

Distance Education in the Future: The Faculty Viewpoint

University faculty members were also surveyed via e-mail. Faculty themes were divergent. Categories included (1) theory to practice, (2) global instruction, (3) comparisons between distance education and face-to-face instruction, (4) cost versus benefit in terms of quality, (5) the need for interactions, and (6) the need to provide lifelong learning. Two vignettes are included to illustrate faculty viewpoints.

Thought and Reflection: Faculty View

Distance education is most assuredly not a new phenomenon. Nor are the suggested treatments new. Increasingly technologically advanced electronic capacities have literally transformed methodologies associated with distance learning. Even so, the greatest distance that remains in instructional design as well as program planning and design is the classical distance between theory and practice.

My vision of educational systems which address the distance issues of the future are those whose central focus will be paying full attention to narrowing the gaps between espoused theoretic and theories in use. Of main concern in my field of vision are experiential learning programs that occur outside the classroom but at the same time are the central focus of the "in-class" curriculum. Such a process would effectively transform classroom experiences into "field-dependent" learning designs.

Unless and until we actually decrease the "distance" between theory and practice in learning and teaching design, can we really espouse "distance education"? I think not!

Lynn Jones, Associate Professor, Iowa State University

Thought and Reflection: Faculty View

In the next 10 years, I worry that we will see many more attempts to jump on the bandwagon of distance education and prostitute sound educational pedagogy, instructional design, and delivery strategies in doing so. During this time, I foresee that many programs of distance education will continue to be built around one-way "talking-head" approaches with little opportunity or incentive for two-way interactive, problem-solving communication between the instructor and the students.

I envision that the next era of distance education, 20 or more years down the road, will bring about broader, wide-scale acceptance of interactive teaching–learning approaches and greater acceptance of asynchronous instructional designs.

I also envision that we will see both more commonly accepted standards for levels of performance of students by institutions in different countries and a breaking down of geographical and political boundaries in enrolling students in programs of distance education.

I envision that we will continue to see both directed and self-directed instructional designs in place. Studying and learning through programs of distance education using many types of media in both synchronous and asynchronous settings is in our future and is here to stay.

James E. Christiansen, Professor, Texas A&M University

Distance Education in the Future: International Training and Development Viewpoint

Professionals in training and development settings focused on themes related to nonformal education and training. These themes were (1) the benefits of reduced travel time and costs associated with training, (2) concerns about the digital divide, and (3) the need for learning centers with mentors. Reaching the needs of learners efficiently is an overall goal. Based on our survey of colleagues in countries such as South Africa, China, Greece, and Turkey, distance

Thought and Reflection: Training and Development View

Worldwide distance education programs are playing a more and more important role in the educational arena. In such populous developing countries as China where on-campus teaching and learning resources are always limited for the whole target-learner population in the society, carrying out distance education programs is an extremely meaningful educational strategy. In recent years, with the rapid development of information and electronic technology, distance education programs keep on updating themselves by adopting the latest computer, Internet, and multimedia-based learning tools to reach more and more distance learners in a more convenient way. It is foreseeable that computer and Web-based distance teaching and learning will be an outstanding characteristic in educational style in the 21st century. Such trends will be seen in China, too.

What does such change mean to China? It means both opportunities and challenges definitely! As to opportunities, such distance education programs would make sharing educational resources in a broader, deeper, and cheaper way become possible among a huge population. Also, changed information exchange style will alter people's minds about the distance to other people, to other societies, to other careers, and finally to the whole world. In another words, it will accelerate the pace of globalization and change people's attitude about distance.

As to the challenges, compared with developed and less populous countries, China has some unique problems to solve in order to utilize distance education programs in a better way. Regional differences in educational resources allocation and development pace are obvious in China. At the same time, lack of fundamental infrastructure and personnel training programs are a relatively common concern for Chinese educators. That means more efforts in these areas are needed in China to help develop sustainable developing power in distance education programs in the near future.

Dr. Yan Li, Assistant Professor, Zhejang University, China

education is a vehicle to reach people in rural areas and improve their well-being and sustainability.

The major themes from each of the perspectives were similar to some of the trends noted from distance education professionals in their books and articles. None of us knows for sure what the future holds, but we think it is evident that distance education is here to stay.

Staying Current, Becoming Global

It is clear that the boundaries between first-, second-, and third-world countries are becoming more blurred due to access provided by technology and other factors of globalization. However, access does not always assure meshing of cultures and success.

The convergence of the information age and the technology revolution on shop floors, offices, and boardrooms of the global workplace has changed the nature of work and the roles that professionals play in ensuring the effective performance of organizations, people, and processes. This has led to workplace changes that range from mechanical to computerized, information based to knowledge based, individual based to team based, and hands on to minds on.

As the new world order continues to emerge, how will cultures and economic status drive distance learning? Will broadband access be a necessity? What form of distance learning will the majority of the world adopt?

One thing for sure is that things will change. How can educators, administrators, and learners stay current in this ever-changing field? Wright and Howell (2004) provide 10 efficient research strategies for distance learning to help busy professionals stay current. These strategies include accessing library expertise, using research assistants, reviewing books from your library catalog, journals, databases, and consulting current awareness services, subscription services, distance education Web portals, associations, and listserv/discussion groups. Therefore, we will leave you with a list of sources that we hope you will enjoy and use as a lifelong learner of distance education practices and applications.

 **Internet Connections**

Distance Education Journals

http://www.ajde.com

The *American Journal of Distance Education* is published by Lawrence Erlbaum with a focus on the practice of distance education in the Americas.

http://www.ed.psu.edu/acsde/deos/deosnesw/deonews.asp

DEOSNEWS is a monthly electronic journal published by Pennsylvania State University, promoting distance education scholarship, research, and practice.

http://www.irrodl.org

The *International Review of Research in Open and Distance Learning* is an interactive online journal with an international scope.

http://www.aln.org/publications/jaln/index.asp

The *Journal of Asynchronous Learning Networks* is published by Vanderbilt University.

http://www.cade-aced.ca/en_pub.php

The *Journal of Distance Education* is published by the Canadian Association for Distance Education.

http://www.westga.edu/~distance/jmain11.html

The *Online Journal of Distance Education Administration* is published by the Center for Distance Education, State University of West Georgia.

http://www.aect.org/Publiscations/qrde.htm

The *Quarterly Review of Distance Education* is a refereed journal with articles, briefs, reviews, and editorials.

Internet Connections

Distance Education Subscription Services

http://www.magnapubs.com

The Distance Education Report is published by Magna Publications in a biweekly format.

http://www.distance-educator.com

Daily e-mail updates through The Distance Educator can help you stay current.

Conclusion

This book focused on distance education applications and practices for educators, trainers, and learners. In closing, we share some final thoughts about future trends in relation to the sections of the book. Foundations for instruction and learning at a distance was based on philosophical and cognitive processes that are unlikely to change significantly. That is why they are foundational—they have already stood the test of time. Adult learning theory is a relatively new field of study and deserves more attention in the design and delivery of distance education courses and programs. Distance learners need to be self-directed, and the courses we create must reflect this principle. Based on future demographic trends, learners will include more women, minority, and senior citizens.

For systematic instructional design, we must move away from the instructor-centered instructional models to learner-centered designs. This requires that the role of the instructor shift to that of a facilitator of vicarious interactions between and among learners, instructors, content, and other resources in different settings. The advances in technology will accommodate these interactions and become more affordable and accessible.

Finally, we ponder the future of administrative issues in distance education. Personnel, budgeting, copyright, policy, and program evaluation must adapt to the educational trends and technological advances. We cannot continue simply

 **Internet Connections**

Distance Education Web Portals

http://www.uwex.edu/disted/

This site is maintained by the University of Wisconsin Extension and includes information on keeping current.

http://www.sheeo.org/helinks/disted-links.htm

This link will take you to the State Higher Education Executive Officers Distance Education Resources.

http://www.distance-educator.com/

This site was founded by Farhad Saba at San Diego State University and is very comprehensive.

http://www.adec.edu/online-resources.html

This link takes you to the American Distance Education Consortium site maintained at the University of Nebraska.

http://www.uidaho.edu/evo/distglan.html

This site is linked to the Engineering Outreach at the University of Idaho and includes various useful materials on Distance Education at a Glance.

http://www-wbweb4.worldbank.org/disted/

This site takes you to the Global Distance Educationnet – Teaching and Learning Technology Management Policy and Programs.

to modify existing structure and function. Effective administration of programs of distance education requires creative thinking and problem solving rather than trying to make it fit the traditional model. The ability to change quickly and be resilient is the key to success.

We hope you have enjoyed the book and have found the resources and activities useful. If you completed each of the application exercises, then you are well on your way to creating a distance education lesson with the core competencies required for distance education professionals. These competencies will transfer incrementally to other lessons, courses, and complete programs.

 Application Exercise

After reading this book and reflecting on the content and your own experiences, write your own vignette about your vision of the future of distance education.

References

Castro, A.P. (2001). Learning in a digital age: Current and future trends in educational technology. Retrieved July 6, 2004, from *www.geocities.com/apcastro111/conteduc/edutech.htm*

Chute, A., Thompson, M., & Hancock, B. (1999). *The McGraw-Hill handbook of distance learning*. New York: McGraw-Hill.

Draves, W.A. (2000). *Teaching online*. River Falls, WI: LERN Books.

Good, D.B. (1999). Future trends affecting education. Retrieved July 6, 2004, from *www.ecs.org/clearinghouse/13/27/1327.htm*

Howell, S.L., Williams, P.B., & Lindsay, N.K. (2003). Thirty-two trends affecting distance education: An informed foundation for strategic planning. *Online Journal of Distance Learning Administration, 6*(3). Retrieved January 8, 2005 from *www.westga.edu/~distance/ojdla/fall63/howell63.html*

Negroponte, N. (1995). *Being digital*. New York: Knopf.

Snyder, D.P. (2004, January). A look at the future: Is technology the answer to education's long-term staffing problems? *American School Journal*. Retrieved July 6, 2004, from *www.asbj.com/2004/01/0104technology focus.html*

Thomson Course Technology. (2003). Main trends. Retrieved July 6, 2004, from *www.scsite.com/dc2003/trends/main_trends_16_45.htm*

Thornburg, D.D. (1998). Reading the future: Here's what's on hand for technology and education. *Electronic School, June*. Retrieved July 6, 2004, from *www.electronic-school.com/0698f1.html*

Wilson, B.G. (2001). Trends and futures of education: Implications for distance education. Retrieved July 6, 2004, from *http://carbon.cudenver.edu/~bwilson/TrandsAndFutures.html*

Wright, T.C., & Howell, S.L. (2004). Ten efficient research strategies for distance learning. *Online Journal of Distance Education Administration, 7*(1). Retrieved July 28, 2004, from *www.westga.edu/~distance/ojdla/spring71/wright71.html*

Glossary

3-D presentations – representation of objects and backgrounds in a three-dimensional space.

Ability – capacity to perform a certain function.

Achievement motivation – learners' need to accomplish things and solve problems regardless of external factors such as feedback and reward.

ACTIONS mode – acronym for Access, Costs, Teaching functions, Interaction and user friendliness, Organizational issues, Novelty, Speed of course development/adaptation. It was proposed by Tony Bates in 1997.

Active learning – type of learning activity in which learners are engaged and instruction is matched to learners' understanding, level of progression, and interest.

Activity-based interactions – interactions or *inter*activities (i.e., critical thinking, creative thinking, information searching, information sharing, and collaborative problem solving) that involve different types of activities. See Bonk and Reynolds (1997) and Harris (1994a, 1994b, 1994c) for descriptions of several different types.

ADDIE – an instructional design model that stands for Analyze needs, Design instruction, Develop materials, Implement activities, and Evaluate participant progress and instructional effectiveness (Hall, 1997; Powers, 1997).

Adult learning – description of the process by which adults learn, especially as influenced by Malcolm Knowles, who defined various characteristics of adult learners. According to Knowles, adult learners are goal oriented, relevancy oriented, practical, autonomous, self-directed, have prior knowledge and experience, and require respect from their instructors.

Affective – consists of attitudes, motivations, beliefs, emotions, and values.

Affective domain – the affective domain relates to attitudes, feelings, and emotions. The taxonomy in the affective domain was developed to describe levels of commitment in the learning process (from lowest to highest): receiving, responding, valuing, organization, and characterization by value.

ALT tags – refers to the alt attribute of the <img> tag which is generally used to place images on a Web site.

Analog video – video that is stored using film, videotape, television video instruments, or any other noncomputer media.

Analogy – inference that two things are similar in some respect.

Analysis – a systematic approach to problem solving. It is a cognitive skill involving the separation of a complex problem into its constituent components for the purpose of their individual study.

Andragogy – a term coined by Malcolm Knowles to describe the art and science of helping adults learn.

Animation – dynamic images or graphics comprising a number of frames that provide the illusion of continuous motion.

APA – American Psychological Association, the largest association of psychologists worldwide.

Apple – a computer company founded in 1976 by Steven Jobs and Steve Wozniak.

Appletalk – networking protocol developed by Apple to communicate between Apple computer products and other types of computers.

ARCnet – Attached Resource Computer Network, a network that uses token passing. It is relatively a low-speed form of LAN data link technology (2.5 Mbps) developed by Datapoint.

Artificial intelligence – the scientific field devoted to the creation of hardware and software that mimics human intelligence and thought processes.

Assessment rubric – scoring guide, usually numerical, that gives specific criteria to assess student performance on a specific task.

Asynchronous – a two-way communication method that does not happen at the same time.

Attrition rate – rate of shrinkage in size or number; drop-out rate of learners in a distance program.

Audience awareness – the ability to write or present on the appropriate level for the intended audience using correct tone and voice.

Audio tools – radio, audio cassettes, videotapes using the audio track only, telephone, and audio-only conferences.

Audiovisual media – media that includes various audiovisual materials such as posters, paintings, slides, videos, films, audiotapes, and videotapes.

Aural learners – learners who best collect information by listening to others, for instance, in lectures and seminars.

Authentic assessment – a type of assessment that measures the student's ability to perform applied work to demonstrate specific knowledge and skills.

Automaticity – ability to perform a skill or a habitually complex task without the conscious deployment of attention.

Baby Boomers – children born in the post-World War II "baby boom" (1945–1960).

Bandwidth – refers to the amount of data that can be sent in a fixed period of time. This term can also refer to a range within a band of frequencies or wavelengths.

Behavioral anchors – characteristics of core competencies associated with the mastery of content.

Behaviorism – a theory that focuses on observable changes in behavior.

Benchmark – a measurement point used to provide a basis for comparison and which serves as a point of reference. It can refer to a skill that learners must master.

Berne Convention (Berne Union for the Protection of Literary and Artistic Property) – The countries represented in the Convention agreed that they would recognize the copyrights of each other and make any changes to their copyright laws to comply with international standards.

Biological intelligence – the general structure and processes of living organisms inside or related to brain regions and sense organs.

Cable modem – a broadband Internet device that transmits data over cable TV networks.

Chunk – a term that was coined by G.A. Miller in 1956 to describe the memory records that encode a small number of units. This method can be used to collect/group stimuli that are stored in short-term memory.

Cognitive – adjective that describes thinking mechanisms that impact an individual's ability to learn and understand.

Cognitive domain – a mind-/knowledge-based domain of learning. The major idea of the taxonomy is that what learners are expected to know can be arranged in a hierarchy from simple to more complex.

Cognitive information processing (model) – describes fundamental mental operations such as how we perceive and remember information.

Cognitive levels – according to Bloom's taxonomy, cognitive levels are from lowest to highest: knowledge, comprehension, application, analysis, synthesis, and evaluation.

COGSS – a Comparison of Generative/Supplantive Strategy. It was created by Smith and Ragan in 1999 based on Gagné's theory. It helps instructors, instructional designers, and learners determine the balance between instructional strategies and learning strategies based on context, learner, and task variables.

Coherence – the development of a clear thesis and introduction, and the inclusion of well-constructed paragraphs with smooth transitions from one part to another in written and presentation materials.

Collaboration – involvement of two or more individuals on a project.

Commission on New Technological Uses of Copyrighted Works (CONTU) – a committee developed by the U.S. Congress in 1974 to investigate copyright issues. In 1978, CONTU suggested an amendment that would clarify the scope of copyright protection on computer software, which resulted in the Computer Software Copyright Act of 1980.

Communication-based interaction – interaction that allows learners to manipulate electronic tools to complete tasks and participate in other learning events. The interface acts as the point or means of interaction, between the learner and the content, instructor, or other learners.

Compatibility – the degree to which a present (current) act or innovation is perceived as consistent with a past act or experience.

Competencies – the underlying knowledge, understanding, and practical and thinking skills to perform a task.

Competency-based behavioral anchors – performance capabilities needed to demonstrate knowledge, skill, and ability acquisition.

Competency models – descriptions of identified roles, outputs, and competencies of a given profession. Competency models can be used as a tool for recruitment and selection, assessment, content development, coaching, counseling, mentoring, career development, and behavioral benchmarking.

Computer-based interactions – a purpose-based framework that includes five basic functions for computer-based interactions: confirmation, pacing, inquiry, navigation, and elaboration described by Hannifin (1989).

Concept mapping – a concept map is a special form of a web diagram for exploring knowledge and gathering and sharing information. Concept mapping is the strategy implemented to develop a concept map, a special form of diagram consisting of nodes or cells that contain concepts, ideas, and links. The links are generally represented by arrow symbols that describe the direction of the relationships.

Consortium – the association of two or more groups/agencies to share information, resources, or services.

Constructivist – a person who espouses the learning theory that the learner acquires the information or solves problems based on existing learning infrastructures.

Continuum – continuing in time or space without break, interruption, or deviation.

Conventional – formal and traditional.

Copyright – the protection granted to authors and developers while also allowing flexibility for the public to have access to original works.

Corrective feedback – type of feedback that identifies areas that need improvement and provides insights into how to revise assignments.

Counterproductive – adjective describing actions interrupting the achievement of a goal.

Course management tools – interface and management of online courses to provide efficiency for the instructor and ease of use by the learners.

Courseware – the media (text, computer program, or CD-ROM) used for educational purposes.

Crystallized (intelligence) – according to Cattell (1963), "Crystallized ability loads more highly those cognitive performances in which skilled judgment habits have become crystallized (hence its name) as a result of earlier learning application of some prior, more fundamental general ability to these fields" (p. 3). Crystallized implies being firmly established and thus not considered consciously.

Cyclical Redundancy Check (CRC) – four-byte error detection information that is used to verify the integrity of the data.

Decision making – the cognitive process of choosing a course of action among two or more alternatives.

Destination address – provides the unique six-byte ethernet address of the device that is supposed to receive the packet. This address can be a computer, server, printer, telephone (in the case of Voice over IP or VoIP) if the final destination is on the same LAN or the default router if the final destination is on another LAN.

Didactical – instruction that is primarily led by the teacher or trainer; lecture.

Diffusion – the process by which an innovation is communicated through communication channels over time from one person to another in a social system.

Digital – refers to the use of a series of 0s and 1s (binary code) to represent information as opposed to waves.

Digital Millennium Copyright Act – a copyright act with implications for online education: "service providers" are responsible for ensuring that copyright permission is given before materials are accessible from Web sites; the owners of copyrights should control access and reproduction of protected materials; and the service provider should prevent unauthorized access.

Digital video (DV) – video that is stored in a digital format. It generally refers to moving images that can be stored in a computer hard disk.

Distance education – process of delivering instructional resource-sharing opportunities to locations where the learner and the instructor do not physically meet at the same place or time.

Distance learners – learners who are separated from the instructor by geographic distance or by time. The learner–instructor interactions are often supported by communications technology such as television, videotape, computers, or mail.

Domain name servers – the Internet protocol (IP) uses addresses that are unique and globally assigned. The IP addresses are represented by a four-byte address such as 128.194.15.2. Blocks of network addresses are assigned to institutions to be reassigned to individual computers. To prevent end users from having to remember these complex numbers, a domain name system is used.

Dunn and Dunn's model – a model developed in 1974 that is based on learning styles. The model measures the learner's preferred modes for concentration and learning difficult information. The authors' conception takes into account multiple interacting factors including environmental, sociological, and emotional variables.

Egos – the conditional mind. The individual's consciousness and inflated feeling to his or her identity.

E-mail servers – repositories of incoming and outgoing e-mail messages. E-mail servers will accept incoming messages and hold them until the recipient is ready to retrieve them. They also filter traffic, discarding messages with viruses attached and unsolicited messages such as spam mail.

Empiricism – the philosophical view that all knowledge is based on experience.

Episodic memory – memories that are learned and cannot be recalled without also recalling the "episode" and the environment in which they were created.

Epistemology – a branch of philosophy that describes the theory of knowledge.

ESEA – acronym for Elementary and Secondary Education Act that was enacted in 1965. It is also known as the No Child Left Behind Act (NCLB). The main mission of ESEA (NCLB) is to provide guidance and funds to K–12 schools, and provide educational opportunities to all children regardless of their background and abilities.

Evaluation – a systematic process applied to educational programs. It became a widely formalized endeavor in 1965 after President John F. Kennedy signed into law the Elementary and Secondary Education Act (ESEA) mandating a formal evaluation for all federally funded educational programs. It is a systematic process to determine the merit, worth, or value of an object, product, or process.

Excel – spreadsheet and reporting program distributed by Microsoft.

Extrinsic motivation – is derived from learners' responses to forces beyond themselves. Extrinsically motivated people respond to reward, praise, good grades, money, and so forth.

Eyeball cameras – also referred to as PC cameras. They are generally attached to the host PC by a cable and mounted on computer monitors to capture moving and still images. Connectix pioneered the industry, and has held over 75% of the market since 1994. Recently, other large companies, such as Kodak, Intel, and Philips, and other smaller companies started contributing to the manufacturing and promoting of PC cameras.

Feature recognition – specific features or components of ideas or concepts are stored and searched for in new stimuli.

Fiber optics – a technology that uses glass or plastic fibers instead of metal cables to transmit data, images, and sound.

Firewall – a specialized computer that is installed between the institutional LAN and the Internet. Depending on its configuration, it can be used to filter inappropriate information such as sexually explicit Web pages, it can be used to protect the institutional LAN from viruses and other harmful traffic, and it can be used to prevent computers from within the institution from being used to launch attacks on the rest of the Internet were a virus to infiltrate the security.

Flash – a software/Internet application created by Macromedia for the purpose of publishing animations on the World Wide Web.

Font – an array of glyphs. It is associated with a set of parameters such as size, posture, weight, and style. Examples of font types include Times New Roman, Arial, Arabic Transparent, and Book Antiqua.

Formative assessment – the process by which information is gathered either through instructor observation, assessment of student work, or various class activities, and used to adapt teaching and learning practices to satisfy student needs.

Frame type or length – the section of the packet that specifies either the type of data that is contained within the frame or the length of the frame.

Functional evaluation approach – evaluation related to the technical and design area of a program, course, or activity and is associated with equipment requirements and specifications. The quality of the program's outcomes cannot be achieved in isolation from functional and managerial levels of quality.

Gagné's Nine Events of Instruction – nine instructional components that can improve instructional planning and engagement to optimize learning.

Gen-Xers – children who were born in the 1960s and 1970s.

Globalization – a rapidly increasing social, cultural, political, and economic process of awareness, though not necessarily acceptance, of a global consciousness and interdependence by which people make decisions about their life, their work, and their physical environment; decisions affected or influenced by expansion and interconnectedness of linkages throughout the whole world, not just the region or country in which they live and work; and decisions that over time collectively result in social, cultural, political, economic, and environmental consequences, both intended and unintended (Christiansen, 2002).

Graphics – images, either photographs or artwork other than text generated by data fed into a computer via a keyboard or a database.

Grasha-Reichman's student learning styles scale – a social interaction scale with patterns of preferred styles for interacting with instructors and fellow learners in a learning environment.

Gregorc's learning style delineator – a measure of bidimensional patterns of learning preferences for making sense of the world through the perception and ordering of incoming information.

Grow's hierarchical stages of learning model – this model includes four stages of self-directedness that can impact instructional design and delivery of distance education.

Hallmark – distinctive feature or characteristic.

Haptic learners – learners who best collect information by feeling, holding, or touching things.

Heterogeneity – diversity and variety in traits or characteristics; different.

Homogeneous – unvaried or undiversified in traits or characteristics; similar.

Hill's cognitive style mapping – a diagnostic technique for determining an individual's cognitive styles. It was developed by Joseph E. Hill, president of Oakland Community College in Bloomfield Hills, Michigan, during the 1960s and 1970s.

Honey and Mumford's learning style questionnaire – Honey and Mumford (1982) expanded upon Lewin's cycle of adult learning which stipulated that adults move through learning in stages: by engaging in a "real-world" experience, by reflecting on this experience, by making general rules about this experience, and then by experimenting with a slightly modified event reflecting the first experience.

HTML – Hypertext Markup Language is the coding language used to create hypertext documents on the Web.

Hub or Switch – a nonintelligent device that sends all information input into any one port or to all other ports on a computer, server, or other information collecting device.

Human resource development – an organized learning experience that involves planning, administering, or evaluating programs designed to enhance competence and professionalism.

Hybrid courses – a blend of online learning with face-to-face teaching.

Hyperlinks o a graphic or piece of text on a Web page that is linked to another Web page, either on the same or on another Web site.

IBM – acronym for International Business Machines Corporation. It develops hardware, operating systems, and applications that support Oracle.

Icebreakers – activities that are used to help learners get acquainted in new subjects or situations.

Ideology – a set of beliefs that form a group or theory.

IEEE – Institute of Electrical and Electronics Engineers known for developing *standards* for the *computer* and electronics industry.

Individualization – discrimination of an individual from a generic group.

Inductive reasoning – process of reasoning and drawing conclusions based on prior observations.

Information technologies – the forms of technology that enables one to design, develop, install, store, transmit, implement, and manipulate information using computer and telecommunication systems.

Innovation decision process – process through which an individual (or other decision-making unit) passes from knowing, to forming an attitude about, to making a decision to adopt or reject, to confirm a decision, or finalize a thought.

Instant Messaging (IM) – Web-based service that combines the live nature of chat rooms with the direct contact of e-mail.

Instructional design – a systematic process or organized procedure for developing instructional materials including steps of analysis, designing, developing, implementing, and evaluating.

Instructional method – a strategy the instructor uses to accomplish an instructional objective.

Instructor centered – the instructor determines the content and organization of the course, activity, or process.

Interactive learners – learners who are actively engaged at different levels in the instructional and learning process.

Interactive video – the use of two-way audio and video for conferencing and instructional purposes.

Interface – a linkage, a shared boundary usually between a computer and a user, or between a computer and a communication medium. Interface with a user usually refers to the components of computer and software that can be manipulated by the user, such as the screens, icons, menus, and dialogues.

Interlacing – a progressive display of graphic images commonly used in televisions and computer display screens to minimize choppiness and to smooth transitions.

Internet Protocol (IP) – the telecommunication protocol used to route a data packet from its source to its destination based on address information carried in the message.

Intrinsic motivation – motivation that is derived from learners' internal drive. Intrinsically motivated people want to learn for the sake of learning and curiosity.

ITU – International Telecommunications Union is an international organization within the United Nations system that coordinates global telecommunications networks.

JPEG – acronym for Joint Photographic Experts Group, a type of file format that uses a lossy compression technique to store for bitmap images to make the file size smaller for transmission.

Key informant – a person whose testimony or description of what exists in or for the client population is available and credible. Program managers often solicit information from such an individual so as to have input to assist in making decisions.

Kiersey temperament sorter – a personality assessment tool that determines the temperament of an individual based on his or her inclinations.

Kinesthetic learners – learners who best collect information by moving or touching.

Knowledge – a body of information applied directly to the performance of a given activity. Also, the condition of knowing facts, processes, or procedures gained through association or experience.

Kolb's learning style inventory (1976) – it ranks strengths and weaknesses in four abilities—Concrete Experience (CE), Reflective Observation (RO), Abstract Conceptualization (AC), and Active Experimentation (AE).

Layout – the physical arrangement of text blocks, headlines, subheadlines, or body copy of any printed or Web-based material.

Learner centered – description of organization and content of instruction largely determined by the learner's needs and perceptions with the instructor's role mainly as that of a facilitator or a coach.

Learner-centered approaches – different techniques and methods that focus on individual learners and their needs to make decisions about how the learning process should occur and how it can be enhanced.

Learner-centered instructional environment – a teaching environment that takes into account the learners' backgrounds, personal beliefs, experiences, perspectives, capacities, and how the instructional methods need to be manipulated to promote high standards of learning and achievement for all learners.

Learner-content interactions – process of interacting with content to affect the learner's understanding, perspective, or cognitive structures. Examples of learner-to-content interactions are online books, online instructional materials, support materials, worksheets, and case studies.

Learner-environment interactions – interactions that occur when learners manipulate tools, equipment, or other objects outside of the computer interface.

Learner-human and learner nonhuman interactions – those interplays or back-and-forth actions and activities that occur between and among the learner and other learners, instructors, resource personnel, and so forth, and the content resources they provide to enhance instruction.

Learner-instruction interactions – interactions that consist of a series of events that are necessary to achieve a defined set of objectives to learn a specified content area.

Learner-instructor interactions – student–teacher interactions undertaken to attempt to motivate and stimulate the learner and to allow for the clarification of misunderstandings by the learner in regard to the content. Examples of learner-to-instructor interactions are lecture, e-mail, online editing and feedback, evaluation of learning, ITV, streaming video, and voice over PowerPoints.

Learner-learner interactions – type of interaction that occurs between one learner and another learner, alone or in group settings, with or without the real-time presence of an instructor. Examples of learner-to-learner interactions are online chats, threaded discussion, e-mail, point-to-point video conference, and audio calls.

Learner-self interactions – interactions that occur within the mind of the learner. They include both the cognitive operations that constitute learning as well as metacognitive processes that help individuals monitor and regulate their learning.

Learning styles or preferences – tendencies of the learner to prefer to process information in different ways.

Learner-technology interactions – examples of learner-to-technology interactions include online tutorials on how to use educational technology, getting help online, downloading plug-ins, installing software, file management including uploading and downloading files, and electronic libraries.

Learning theory – a theory that attempts to explain and predict behavior during the learning process.

Legitimate peripheral participation – an activity that provides a way to speak about the relations among newcomers and old-timers, and about activities, identities, artifacts, and communities of knowledge and practice.

Local Area Networks (LANs) – information networks that are typically built within a building or contiguous buildings and run at high to very high speeds (billions of bits per second up to 10 billion bits per second). They are made up of two types of devices: the first could be considered infrastructure devices and the second, network servers. The infrastructure devices consist of the network interface cards that connect the hosts to the network, the wire plant, the switches, and the routers. The servers each provide a specific network function such as file or print serving, domain name resolution, firewall, rate shaping, network management, and address assignment.

Long-term memory – the portion of memory that is relatively permanent.

Lower-level thinking – information processing at the lowest levels of Bloom's taxonomy (knowledge, comprehension, and application).

Management – the action of planning, organizing, coordinating, running, directing, coordinating, controlling, and evaluating the use of people and materials to accomplish the goals of an organization.

Managerial evaluation approach – related to how successfully the relationships within and outside the parent organization are fostered and managed as they relate to the distance education mission of the organization.

Mediator – intermediator, go-between to link people; can be an ombudsman in some cases.

Meta-analysis – a method that combines the results of several studies to integrate their findings.

Metacognitive – adjective describing learners' automatic awareness of their own thoughts and cognition in the process of metacognition.

Metadata – information, data about data.

Metaphor – a figure of speech in which an expression is used to refer to something in terms of another.

Millennials – people born in the 1980s.

Mnemonic – the use of letters in a sequence to facilitate storage and recall.

Modem (modulator-**dem**odulator) – a device that enables a computer to convey data over telephone or cable lines.

Monochrome – having only one color (text or background). Most frequently applied to black-and-white photographs but can also apply to a computer screen that displays information in only one color on a black or dark background.

Motivation dimensions in instructional theory – when a learner is engaged in the instruction because he or she is interested, finds the material to be relevant, expects to apply the information, and finds it satisfying.

MPEG (Moving Pictures Expert Group) – an international standard for a compression algorithm for video/audio files.

Multilevel evaluation approach – use of evaluation techniques that include more than one level, such as functional, managerial, and instructional.

Multimedia – refers to bringing together a number of diverse technologies of visual and audio media for the purpose of communicating. Examples of multimedia include text, graphics, audio, video, animations, and simulations.

Municipal Area Networks (MANs) – typically fiber optic or wireless-based networks that connect two or more buildings within a community to each other at high and very high speed.

Myriad – countless, multitudinous.

Myth – tradition or fable; popular unfounded belief.

Negative reinforcement – the removal of a consequence that the learner found rewarding.

Network Management Servers – devices that monitor the performance of networks, report changes in operational status, and support security, such as end-user authentication.

Novice – someone who is new to an activity.

Object-oriented software – an object considered a **"black box"** that receives and sends *messages*. A black box contains sequences of computer instructions (code) and information on which to operate (data).

Observability – the degree to which the results of an innovation are apparent to other people.

Olfactory learners – learners who process information best by tying it to smells or tastes.

One-way live video – demonstrated by programs downloaded by a satellite receiver to an audience or individual not able to participate by being on location with the speaker or instructor. In this case, the learner would participate as a passive listener similar to watching a television program.

One-way live video with two-way audio – a delivery technology using one-way live video with phone, Web, or -mail/chat features for participant interaction.

Openers – methods that are used to introduce participants to the content at the outset of the learning experience.

Operant – a type of response that produces an effect or has an influence on the environment.

Overt – apparent; observable.

Pattern – the process whereby environmental stimuli are recognized as examples of concepts and principles already in memory.

Payload – the actual data to be delivered through a network. This section of the packet can vary between 46 and 1,500 bytes in length.

Pedagogy – the art and science of teaching children and applying educational theory to enhance their learning.

Philosophy of education – a philosophical study or view of education. Major schools of thought include liberal, progressive, behaviorist, humanist, radical, and analytical philosophies.

Portfolio – a collection of work that exhibits the learners' process, progress, and achievements.

Preconceptual activity – activity where no conscious thought is involved.

Primary distance learners – learners who are not actually tied to a campus or learning facility but who are motivated by a need to have access to a continuing education or formal program for personal advancement.

Principles of good practice – guiding principles for the field of distance education that improve support, instruction, and learning.

Problem solving – a systematic process of identifying, analyzing, and evaluating a particular problem to arrive at workable solutions.

Prototype – a standard or typical example used as a reference for later work.

Psychomotor domain – learning domain that refers to the use of basic motor skills and physical movement.

Psychomotor levels – domain described with six different levels where the lowest level is represented by simple reflexes and the highest level is represented by more varied neuromuscular coordination.

Quantum-bit computing – a classic computer based on quantum theory. Quantum theory explains the nature and behavior of matter and energy on the atomic and subatomic level.

Reflection – a metacognitive activity in which the learner thinks about and organizes information from various learning activities, such as reading and discussion.

Relative advantage – the degree to which a new idea, process, or product is perceived as being better than a previous idea, process, or product.

Resolution – the number of pixels per square inch creating a visual image; the higher the resolution, the clearer the image.

Rigor – level of hardship, severity, difficulty.

Search engines – databases of Web sites that help users to search the Internet for other Web sites based on keywords or sentences. Examples of search engines include Google, Infoseek, Lycos, Excite, and Altavista.

Selective attention – refers to the learner's ability to direct attention in specific directions.

Self – an individual's inner or mental consciousness.

Self-directed learning – the ability of learners to direct their own learning.

Session-initiated protocol (SIP) – the session spent on a Web site by a user with a unique IP address.

Shaping – the reinforcement of successive approximations to a goal behavior.

Simulation – a series of photographs, drawings, videos, or sound recordings creating the impression of a virtual experience.

Skill – an observable competence to perform a learned act that requires motor ability.

Stakeholder – any groups or individuals within or outside the organization who have an interest in the activities and performances within the organization.

Streaming video – sequence of moving images delivered in packets rather than as a complete download over the Web.

Student-centered, technology-rich learning environment (SCenTRLE) model – a model with eight instructional events for facilitating construction of knowledge and the development of metacognitive skills associated with lifelong learning.

Survey – the act of gathering and studying information to improve comprehension or analysis of a subject matter.

Synchronous – two-way communication that is simultaneous or occurs at the same time. Examples of synchronous methods are Internet chat rooms and desktop videoconferencing systems.

Synthesis – refers to the combination of ideas, facts, or elements to form a new whole.

Systematic instructional design – the process of translating general principles of learning and instruction into plans for teaching and learning.

Taxonomy – a classification system to order concepts or ideas.

Technology, Education, and Copyright Harmonization Act (TEACH Act) – copyright act that expands categories performed in distance education to reasonable and limited portions, recognizes that a learner should be able to access the digital content of a course wherever he or she has access to a computer, allows storage of copyrighted materials on a server to permit asynchronous performances and displays, permits institutions to digitize works to use in distance education, and clarifies that participants in authorized distance education courses and programs are not liable for infringement.

Template matching – an exact mental copy of the stimulus is stored in memory.

Tool-based interactions – five levels of interactions based upon telecommunication tools, such as electronic mail and delayed messaging, remote access and delayed collaboration, real-time brainstorming and conversation, real-time text collaboration, and real-time multimedia and/or hypermedia collaboration.

Traditional correspondence – form of written instructional materials sent or received by traditional mail.

Transactional distance – a measure of distance as a pedagogical phenomenon. It involves the interactions between and among the instructors, the learners, the content, and the learning environment

Transcript – a text representation of sounds in a media representation or an auditory tract.

Transmission Control Protocol (TCP) – a major part of *TCP/IP* networks. *IP* protocol deals only with *packets* whereas TCP protocol enables two *hosts* to establish a connection and exchange streams of data.

Trends – tendencies, general directions, overall movements.

Trialability – the degree to which an innovation may be experimented with on a limited basis before making a decision to adopt or reject.

Two-way live audio and two-way live video (compressed video or ITV) – allows for the transmission of synchronous video and audio between two or more sites. A computer, called a "CODEC," compresses the video and audio signals so that they can be transmitted using lower bandwidth.

Universal design – designing materials that everyone can access. The concept of universal design has been used in the field of architecture for generations, yet it has become important in education and training in the past decade with the integration of technology.

Vicarious interaction – the intersection of various interactions that promotes or maximizes learning.

Video – electronic recording and playback of imagery in a television system.

Videoconference – the act of conducting a conference or a meeting using video and audio signals to link participants at different and remote locations.

Virtual reality – use of computers to simulate a real or imagined environment that appears as a three-dimensional (3-D) space.

Visual learners – learners who collect information best when looking at charts, diagrams, pictures, and observing other people at work.

Western Association of Schools and Colleges – a regional accrediting association that is composed of three commissions: Accrediting Commission for Senior Colleges and Universities, Accrediting Commission for Schools, and Accrediting Commission for Community and Junior Colleges.

Western Interstate Commission for Higher Education (WICHE) – this commission was established by the Western Regional Education Compact in the 1950s for the purpose of facilitating resource sharing among its member states and has published a number of documents including *Good Practices in Distance Education* (WCET, 1997).

Wide Area Networks (WANs) – networks that connect LANs and MANs to each other over much longer distances, usually using circuits provided by carriers. These circuits can be carried over fiber optic cable, microwave, wireless, or satellite facilities.

Wire plant – consists of four-pair twisted pair wires between the offices or classrooms and the wiring closets. Wiring closets are typically connected to each other using a fiber optic cable which supports ethernet connections at long distances (typically up to 550 m for 1 billion bit per second connections and up to 2,500 meters at lower speeds).

World Wide Web (WWW) – a network of information that includes text, graphic, sound, and moving images. A collection of all the resources accessible on the Internet mainly via HTTP or via older protocols and mechanisms, such as FTP or Gopher.

About the Authors

Kim E. Dooley (k-dooley@tamu.edu) is an associate professor and chair of the distance education workgroup in the Department of Agricultural Education at Texas A&M University. She received her PhD from Texas A&M University. She has conducted numerous professional presentations and training programs around the world. Her research focuses on learner-centered instruction design and delivery strategies in the context of distance education. Her scholarship is communicated through more than 100 publications including several chapters focusing on distance education. She has received numerous honors and awards for scholarship, teaching, and program development. Within the department, Dr. Dooley is helping to develop and deliver the Master's of Agriculture at a distance program and the doctoral Doc@Distance program.

James R. Lindner (j-lindner@tamu.edu) is an associate professor and chair of technology-assisted learning in the Department of Agricultural Education at Texas A&M University. He received his PhD from Ohio State University. He has established a national reputation as a rigorous scholar and prolific author focusing on planning and needs assessment, and research, measurement, and analysis in the context of distance education. Within the department, Dr. Lindner is helping to develop and deliver the Master's of Agriculture at a distance program and the doctoral Doc@Distance program. He has authored or coauthored two books, several chapters, and more than 150 articles and papers. He has received numerous honors and awards for presentations of research findings at international and national conferences.

Larry M. Dooley (ldooley@coe.tamu.edu) is an associate professor and chair of the human resource development program in the Department of Educational Administration and Human Resource Development at Texas A&M University. He received his PhD from Texas A&M University. Dr. Dooley is president of the Board of the Academy of Human Resource Development (AHRD) and serves on the Board of the Academy of Human Resource Development Foundation. His research interests are in the integration of distance learning into organizations for performance improvement, leadership development, and foundations of human resource development. He has authored more than 100 publications appearing in professional journals and books. He is also founder of The TECH Training Group, Inc., an international consulting firm.

*   *   *

Chehrazade Aboukinane is a PhD student in the Department of Agricultural Education at Texas A&M University. Her research focuses on educational technology and delivery strategies. She received her BS in biological systems engineering and MS in agricultural engineering from Texas A&M University. Ms. Aboukinane was nominated as a future leader in CIGR (International Commission of Agricultural Engineers) and has presented papers at professional meetings.

Rhonda D. Blackburn is the lead IT consultant at Texas A&M University's Instructional Technology Services. She develops workshops that help faculty incorporate technology into their courses. During the workshops she emphasizes planning, instructional design, and andragogical practices. Along with training, Dr. Blackburn teaches psychology while using the same principles and practice she stresses to the faculty.

Barry L. Boyd is an assistant professor in the Department of Agricultural Education at Texas A&M University. He earned his doctorate from Texas A&M in 1991 with an emphasis in leadership education and instructional design. His research in learner-centered instructional methods has been published in six refereed journals. His emphasis as a student-centered instructor has earned him recognition from both students and faculty. Dr. Boyd consistently ranks in the top 25% of the teaching faculty at Texas A&M and was recognized in 2003 with the Association of Former Students College Teaching Award.

James Buford, Jr. is the director of Ellis-Harper Management, a human resource consulting firm, and adjunct professor of management at Auburn University. He also serves on the faculty of Troy State University-University College and is affiliated with the Center of Local Government Studies. He earned his BSF and MS degrees from Auburn University and his PhD from the University of Georgia. Dr. Buford is certified by both HRCI and PHRCC. Dr. Buford has authored and co-authored three texts and more than 75 articles and papers. His work has appeared in many of the leading HR management journals. In addition to his research, teaching, and continuing education activities, he has consulted widely with organizations in the areas of selection, performance appraisal, and compensation. He was involved in the development of one of the first accreditation programs in the United States for local government HR professionals. He also writes creative nonfiction, including two critically acclaimed collections of essays.

Yakut Gazi is a PhD candidate in the Department of Educational Psychology at Texas A&M University, specializing in educational technology. She received her BS and MA at Bogazici University in Istanbul, Turkey. She taught undergraduate students for several years and is currently working as a multimedia specialist. She is interested in how people communicate in intercultural online environments and how this communication is affected and shaped by the technological features of the communication platform as well as people's backgrounds.

Atsusi Hirumi is an associate professor of instructional technology at the University of Central Florida. Born in New York, Dr. Hirumi spent most of his formative years growing up in Nairobi, Kenya, East Africa. He received his BS in biology from Purdue University with a secondary teacher certification in biology and general science. He received his MA in educational technology from San Diego State University and his PhD in instructional systems from Florida State University. Dr. Hirumi's work focuses on developing systems to train and empower K12, university and corporate educators on the design, development and delivery of interactive distance education programs. His research concentrates on the design and sequencing of e-learning interactions.

He has published over a dozen articles, several book chapters, and has made over 100 presentations at international, national and state conference on related topics.

Kathleen D. Kelsey earned her PhD from Cornell University in 1999. Her dissertation focused on interaction in a distance education setting. She received the Alan A. Kahler Outstanding Dissertation Award from the American Association for Agricultural Education for this work. She has continued to research distance education issues while teaching research and evaluation methods at Oklahoma State University. Evaluation of educational programs has become Dr. Kelsey's main research focus. Dr. Kelsey is a member of the American Evaluation Association and adheres to the Guiding Principles for Evaluators and follows the Program Evaluation Standards advocated by the AEA.

Walt Magnussen has his Bachelors and Masters degrees from the University of Minnesota and his PhD from Texas A&M University. He has been teaching continuing education classes in data communications since 1985. He is currently the director for telecommunications for Texas A&M University where he is responsible for 35,000 telephone subscribers, several thousand cable television connections, all wireless communications, fiber optic cable plant and a wide area network that connects close to 200 sites both in the State of Texas and internationally. He is also the director of the Internet2 Technology Evaluation Center, one of four centers that promote the deployment of advanced technologies to Internet2 member universities. Other responsibilities include chairing the Operational Subcommittee of the Telecommunications Planning and Operations Council (legislatively mandated to oversee State Telecommunications in the State of Texas), chairing the publications committee for the Association for Communications Technology Professionals in Higher Education (ACUTA) and Co-Chairing the Voice over IP Working Group for Internet2. Dr. Magnussen has worked in over 30 countries in establishing Distance Education Networks. He helped in the design of the Lonestar Education and Research Network (LEARN), one of the largest state optical backbones in the USA.

Tim Murphy is associate professor and assistant department head for Graduate Programs in Agricultural Education. He advises graduate and undergraduate students, teaches courses in methods of teaching and technological change, and conducts research in methods of technology-assisted learning and distance education.

Susan Eugenia Wilson is a graduate assistant and student in the Department of Agricultural Education at Texas A&M University. She studies conceptual and concrete connections between communication and education. She earned her bachelor's in agricultural journalism from Texas A&M University in 2003.

Index

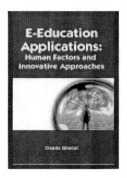

Breinigsville, PA USA
28 November 2010
249962BV00002B/1/P